GOING NORTH THINKING WEST

GOING NORTH THINKING WEST

The Intersections of Social Class, Critical Thinking, and Politicized Writing Instruction

IRVIN PECKHAM

UTAH STATE UNIVERSITY PRESS
Logan, Utah
2010

Utah State University Press
Logan, Utah 84322-3078

Manufactured in the United States of America
Cover design by Barbara Yale-Read
ISBN: 978-0-87421-804-6 (paper)
ISBN: 978-0-87421-805-3 (e-book)

Library of Congress Cataloging-in-Publication Data

Peckham, Irvin.
Going north, thinking west : the intersections of social class, critical thinking, and politicized writing instruction / Irvin Peckham.
p. cm.
Includes index.
ISBN 978-0-87421-804-6 (pbk.) — ISBN 978-0-87421-805-3 (e-book)
1. English language—Rhetoric—Study and teaching (Higher)—Social aspects—United States. 2. English language—Rhetoric—Study and teaching (Higher)—Political aspects—United States. 3. Critical thinking—Study and teaching (Higher)—United States. 4. Working class—Education (Higher)—United States. 5. Social classes—United States. I. Title.
PE1404.P383 2010
808'.0420711073—dc22

2010021911

CONTENTS

ACKNOWLEDGMENTS

I would like to dedicate *Going North* to those who have made it and my life as a writing teacher possible. First was my mother, who is now slipping away. She gave me the gift of literacy when she brought home from the library a book about Buffalo Bill instead of the comics I had requested to fill my days while I was staying home from school with the mumps. I hid my disappointment. I opened the book the next day after my mother and father had left for work, inwardly vowing not to like what I was seeing, but after a few pages, I was off and running. The books piled up.

Next is Charles Cooper, who patiently initiated me into the academy. As this book will make clear, I have a certain degree of working-class contempt for the academic ethos, but I have loved my life as a writing teacher, working with students and learning with them about the part writing plays in the drama of the self struggling like Michelangelo's St. Matthew to free itself from the rock.

Equally as important are my friends in the rhetoric and composition community, where I have found my third home. I have treasured our friendships in our virtual social life that momentarily materializes when we see each other in conferences before returning home. I have known them in varying degrees of intensity and for different durations. Collectively, they have given a deep color to the tapestry of my professional life in a way I can't begin to explain. I would like to name them, but when I call someone in, I leave others out, which would contradict what I mean.

By far most important have been my wife, Sarah, and my two children, Heather and Jesse. All of you know what family members have to put up with when one of them decides he or she has something to write.

1
INTERSECTIONS

This book is about teaching, which is far more than a simple act of transmitting knowledge from those who know into those who are learning—or even of initiating the young into our matrix of discourse communities. The classroom is where community happens, the site of cultural reproduction and revolution, of parroting and creating, of being and not-being. It is the site of power struggles between social classes through the agency of language, where we sort students and distribute privileges, where we train students to accept the kind of life they will most likely have as adults. It is also the place where we were trained to be teachers and where we are constantly being retrained through our praxis. Looked at this way, the classroom is a very interesting place.

This book is also about writing. Writing is a fundamental act of literacy, of naming the world and writing one's way into it. But writing, like teaching, is far from simple. Words, which form the fabric of writing, remove us from primary experience. They shape our understanding and identities. In a literate society, words are a primary agency of exchange. Words, even more than weapons, are consequently the tools of power. Words form webs of aggression and deceit. Through words, we sort people, create and maintain hierarchies, and distribute privilege. They are the way we do things—and they are the agency through which things are done to us. They are the vortex of culture, which is why words and literacies are also very interesting.

My interest in the intersections between writing, teaching, and social class is personal because I have changed my identities, allegiances, and ways of thinking as a consequence of my career. I was born into a rural, working-class family. I am now urban and excessively middle-class, although ineluctably carrying my working-class origins with me. I began working as a high school teacher; I now teach, research, and write in a doctoral intensive university. I used to think that educational institutions functioned to encourage students to learn. I now see them as functioning in part to create failure. The primary agency of failure is language. Although this conflicted interpretation of education may be obvious for postmodern language theorists with a liberatory bent, it is a far cry from what I imagined when I was a high school student, a college student, or an idealistic high school teacher who imagined writing as the road to satori.

I also have to consider that I am an agent of the social reproduction project institutionalizing failure for the majority of working-class students, a disproportionate percentage of whom are women, African-Americans, and Hispanics (Zweig 2000, 31-33). What's worse for me, writing plays an important role in that project. In some cases, writing leads to knowledge, but in others, it is one of the best ways of sorting people. This is not a particularly surprising claim when one considers some of the common ways in which we use language to identify ourselves and the social groups to which we belong. Consequently, an analysis of writing instruction as a sorting mechanism bends back to what I do. Although I do not consider myself to be a radical writing teacher hell-bent on restructuring society through the agency of my students, my sympathies lie in this direction, so I am particularly concerned about classroom strategies that I and other progressive teachers might be employing that contribute to the reproduction of social class relationships in spite of our intent to challenge them.

Marxist in origin, social reproduction theory has been expanded, refracted, and complicated by Althusser (1984), Gramsci (1971), Durkheim, and more lately, Freire ([1970] 1995), Berger and Luckmann (1967), Bowles and Gintis (1976), Bourdieu and Passeron (1990), and a host of other writers in the 1980s, 1990s, and 2000s (Shor 1980; Anyon 1980;Apple 1982; Katz 1971; Clark 1960; Aronowitz 1997, Gee 1997; Macedo 1993; McLaren 1989; Giroux 1993, Berlin 1987, to name some of the more widely cited). The foundation of social reproduction theory seems fairly commonsensical. Societies are framed by social structures that define and maintain the relationships among various groups within the society. Among industrialized societies, in particular, these relationships are characterized by a hierarchical distribution of wealth, status, and privilege. Some groups at the top of the hierarchy receive significantly more wealth, status, and privilege than those at the bottom. In the United States, this difference is bracketed by corporate executives whose average salaries (including stocks and other compensations) are about 10.9 million dollars a year (AFL-CIO 2009) compared to a minimum wage earner who makes about 14,000 dollars a year. For a variety of reasons and through different agencies, societies tend to reproduce themselves through myths, social relationships, and institutions such that the people who have the most keep the most—and in fact end up getting even more until the obscenity of the disparity sparks reconstruction or revolution. It stands to reason that those groups who have the most also have the most to say about how the society gets reproduced through legislation, leadership, religious and educational institutions—as Marx (1846) said, "The ideas of the ruling class are in every epoch the ruling ideas." It also stands to reason that they will like social structures the way they are and will work to keep them that way.

Of course social structures aren't quite that simple. Thus, we have resistance theory. The roots of resistance theory lie in Gramsci's *Prison Notebooks* (1971), in which he details the necessity for legitimating inequality through institutions, intellectuals, and politicians. The necessity for this legitimation lies in the contradictions contained within the democratic, capitalist system, such as when we pray to Christ and Carnegie in the same breath. Although obscured by ideological state apparatus (Althusser 1992), people intuit these contradictions; consequently, the social system, dominated by those who benefit most from the way things are, creates a series of narratives naturalizing the inequality and suppressing the contradictions. But not everyone goes along with the game. Pockets of resistance develop on both an individual and group level. Thus, there is constant friction within the social system. This friction leads to resistance and either change or repression.

Resistance theory in education has been present in educational narratives for some time; it was implicit in Charles Dickens' *Hard Times* ([1854] 1958) or Edward Eggleston's *The Hoosier Schoolmaster* (1913). It is in the background of Michael Katz's (1975) reconstruction of the growth of mass education in the United States. It is more fully theorized in Paul Willis' (1977) ethnography of working-class students. Willis narrates the process and consequences of "the lads'" subversion of classroom instruction. We see classic strategies like dropping pencils, whispering, making fun of the teacher behind his or her back, flouting authority, swaggering, sneering, coming in late, lining up at the door ten minutes before the bell, and work refusal. The subversion works in complicated ways. On the surface level, it obstructs the educational process and severely constrains learning for other students. The lads by and large succeed in their goal of salvaging their integrity in the face of a system that derogates their working-class ethos. The salvaging is perhaps best symbolized by their fundamental goal of "having a laff" (29), generally at the expense of the teacher, who represents authority—i.e., the boss.

On the other hand, the lads play into the hands of the capitalist system that needs laborers. It could be argued—and this is the obverse side of resistance theory—that the lads are unwittingly supporting the overall structure of the system. The system needs failures, because the failures become the laborers. If the system works well, it will internalize in working-class people the responsibility for their own failure. Burton Clark (1960) in his seminal article, "The Cooling Out Function in Higher Education," describes how this kind of internalization works in the community college system.

Michael Apple (1982) in *Education and Power* brought into clear focus the complications that develop from reproduction and resistance theories in education. The lads, for instance, may be successfully educated into failure so that they will take their places as laborers, but they carry into the workplace resisting strategies that they learn in school; these strategies consequently undermine production—as workers, the lads will spend a good deal of time making fun of the boss or, as Stanley Aronowitz (1999) claims, smoking dope in the john (see also the poetry of Jim Daniels, 1985, "Factory Jungle"). So in fact, the social design that internalizes in working-class people the responsibility for their own failure also works against productivity and social coherence.

One has to consider not only the contradictions in the "education" of the lads but also in the education of the "ear 'oles" (the lads' wonderfully descriptive name for good students). Although the ear 'oles may meld into the dominant classes in adulthood, they never lose their working-class origins and the internal conflicts that arise from the past rubbing against the present (see Dews and Law 1995; Ryan and Sackrey 1984; Shepard, McMillan, and Tate 1998; Zandy 1994). As a working-class ear 'ole myself, I have internalized this conflict. All sorts of contradictions follow from working-class people who cross over. The crossing-over works in favor of the capitalist/democratic social structure by bringing into the dominant classes the supposedly more capable, aggressive, and intelligent members of the dominated classes (or at least this is the myth that the democratic capitalist narrative would have us believe and that ear 'oles like myself are only too happy to support); these ear 'oles, coming with perspectives different from the natives' will also have new ideas. To use a self-damning cliché, they might be able to think outside the box because they are never fully in it. And finally, such crossing over reinforces the meritocratic myth of social evolution—i.e., the cream will rise to the top.

On the other hand, the crossing-over works against the dominant class because the ear 'oles retain some allegiance to their origins, or as James Paul Gee (1996) has called it, to their primary Discourse. They may try to disrupt the dominance of the dominant classes. But to turn the table once more, by disrupting the old order, cross-overs may in fact be contributing to a future overall social coherence because immutable social structures ossify. Social structures—like languages and people—that maintain a dialectical relationship with new conditions survive. These cross-overs bring with them news of new conditions.

However, crossing over is tricky business. Not many make it—about 1 out of 15 students in the lower income quartile make it through college, compared to about 1 out of 2 from the upper quartile (Zweig 2000, 45). While acknowledging that not everyone starts from the zero yard line, the myth of meritocratic democracy, instantiated in the Horatio Alger narratives,[1] lays the blame for this disparity on the shoulders of those who fail to "rise" above their origins. If students can't do well in school, it's their fault or the fault of their parents, who don't raise them right. This narrative is constantly supported by the media's spotlighting immigrant success stories, like Richard Rodriguez's (1982), Arnold Schwarznegger's, or more generally, comparing mobility statistics of successful immigrant groups to unsuccessful immigrant groups, relying on inductive fallacies to obscure the difficulties that working-class children face in school.[2] In a sense, the American mythology has it right: the home conditions of working-class children *is* why they have such difficulty in school. But the disparity is not because of the inferiority of those

1. Alger was born to a middle-class family complicated by its tilt toward the intellectual fraction. But because of the wealth on his mother's side, Alger was able to go to Harvard. One might speculate that he wrote his dime-store novels featuring Raggedy Dick and the myth of meritocracy just as some latter-day dime-store novelists write steamy narratives that readers buy in Albertsons.

2. I am using working class to describe all social groups who generally work for lower wages, who have very little control over their own labor and time, and possess, in Pierre Bourdieu's (1984) terms, very little symbolic capital. In contrast to the American mythology equating the working class to whites—and in particular to white males, less than half of the working class, according to Michael Zweig (2000) are white males. A significant proportion of the working class is female, black, or Hispanic (33). By including in the working class women and different racial groups, I do not in any way intend to obscure the special conditions of oppression they face.

conditions; it's because of the difference between behavior and values that are taught at home and behavior and values expected in school (see Brice Heath 1983; Lareau 2003). More than difference, the real problem lies in the conflict between the home and school ethos. James Paul Gee (1996) calls this the conflict between the primary and secondary Discourse—by discourse with a capital D, he is referring to all the habits, values, systems of relationships, and traditions through which a social group speaks itself; Peter Berger and Thomas Luckmann (1967) frame the conflict as one of primary and secondary socialization.

The problem lies not only in the conflicts between these two Discourses but also in the frame that privileges the school Discourse and marginalizes the home Discourse of the working classes. In *The Social Construction of Knowledge*, Berger and Luckmann (1967) include a fascinating discussion on general processes involved in moving from primary to secondary socialization—and at a higher level of generalization, from one social group to another. When the primary and secondary worlds are in relative consonance with each other, the child integrates the new world into her former world, re-reading herself in a way that maintains her former notion of the world. The new world may also present an alternative world, that is, the different ways of understanding reality do not cancel the old world, in which case the child is expanding the self, or adding to her old world. But when the new world cancels the old world, the child experiences identity rupture. Although Berger and Luckmann use as an example of radical disruption a Catholic who moves into a non-Catholic world, their description aptly describes any school child who finds herself in a new world that cancels her old. When she first entertains doubts about basic tenets of her primary world, she smiles at herself, internalizing the knowing smiles of others in her primary world. But as the "plausibility structure" of the old becomes less available, "the smile will become forced and is eventually likely to be replaced by a pensive frown" (155-156). Rather than re-read herself, the child has to step outside her old self, risking anomie, a no-self alienated from both worlds.

One thinks of the differences between the home and school selves of the children Shirley Brice Heath (1983) described in *Ways with Words*, her report on a ten-year ethnography of three social groups in the Piedmont Carolinas. With few exceptions, the children from the black working class were energetic, loved, and eager learners in their home communities, but by the time they had (or hadn't) finished primary schools, they had settled into the failure narrative. As Brice Heath describes their school failures (360), one imagines their frowns. What they had learned about *being* at home didn't work at school. Brice Heath's work made explicit the many ways in which the literacy practices of black and white working-class communities were in direct conflict with the literacy expectations of their white, middle-class teachers.

More recently, Annette Lareau in *Unequal Childhoods* (2003) tracked the child-rearing practices of twelve urban families with children in the fourth grade to interpret their consonance with behaviors expected in academic and professional environments. The families were divided along class, race (White versus Black), and gender lines. Lareau's work substantiated (with the caveat that she resists over-generalizing) that working-class and school environments were in marked conflict, whereas the middle-class home environments were essentially prep schools for institutionalized educational settings—and for professional workplaces. With some apologies for the values embedded in the metaphors, she characterized the working-class home environments as dominated by a concept of *natural growth* and the middle-class home environments by a concept of *concerted cultivation*. Lareau found that working-class parents tended to construct borders (like behavioral and geographical fences) around their children within which the children were free to grow; the middle-class parents worked without borders—behavior was subject to negotiation and the children typically traveled frequently and far. Within this generally ambiguous and negotiated environment, middle-class parents prepared their children for middle-class, professional worlds characterized by fluid, ambiguous situations requiring sophisticated negotiating skills through language. The middle-class children were raised to move seamlessly between their private and public worlds. They were made to feel as if they belonged in school and in professional spaces, creating what Bourdieu (1984) called the "entitlement" effect (23). They would assume, for instance, they would *naturally* go on to college and get advanced degrees. That's their birthright. Working-class kids think they're lucky to get to college—and when they do, they know they're strangers.

Brice Heath and Lareau investigated the context producing literacy practices. Other researchers, most notably Basil Bernstein (1971) and Gee (1996), focused on the linguistic and cognitive differences between social groups and linguistic and cognitive expectations in middle-class schools. The problem with this kind of research tracks back to Eric

Havelock's (1963) analysis of the differences between oral and literate traditions, the literate tradition being the privileged one, evidenced by the dominance of literate over oral cultures. The oral/literate binary was reframed in Bernstein's work as restrictive/elaborated linguistic codes. The restrictive codes are linked to narratives, which are in turn linked to oral discourse, characteristic of dominated social groups. The elaborated codes are linked to exposition and written discourse, characteristic of dominating social groups. The different logics operating in the different discourse modes are the logic of time for the simple minded (chronology) and the logic of deduction for the sophisticated thinkers (tautology). As I will argue in chapter five, enforcing this hierarchy of logical modes underwrites the traditional sequencing of required writing programs.

Richard Ohmann (1982) strongly objected to the privileging implied in Bernstein's analysis of the differences between working- and middle-class languages. Ohmann claimed that because of the link between language and cognition, Bernstein's analysis describes in the working classes an inferior way of thinking and that the only way out of their oppressed condition lay in learning (and learning from) the habits of language and thinking characterizing the dominating classes. To be fair, the point of Bernstein's analysis lay in other directions—he was interested in the fit between the working-class ethos and the new style of education with open classrooms and student-centered curricula. His argument was that the new curriculum acted against working-class children in unintended ways.

Similar criticisms have been leveled at Richard Rodriguez's *Hunger of Memory* (1982), Brice Heath (1983), and Mike Rose's *Lives on the Boundary* (1989). In these books, one can interpret the marginalized, oral-oriented Discourses of the dominated social groups as deficient and the literate, middle-class, school Discourse as the goal. John Trimbur (1993a, 44), in his critique of *Lives,* calls this a "reconciliation to the party of order." Elspeth Stuckey (1991) in a memorable critique of the ideology of "literacy" levels a scathing rebuke at all research predicated on the superficially veiled assumption of the superiority of literate culture. She points out that people who possess literacy have everything to gain by "discovering" that literate people have more sophisticated cognitive operations than non-literate people. A series of claims substantiate these discoveries, most of them emanating from the Havelock (1963) and Ong ([1982] 2000) traditions. These claims are ones to which

English teachers, in particular, are drawn, for they venerate (drum roll) *The Word.* In our English departments and perhaps among legitimated intellectuals[3] in general, one frequently hears hymns of the written page. Stuckey makes clear the irony of egalitarian intellectuals perpetuating through their own literacy practices the social structures they imagine they are undermining.

Stuckey (1991) claims that teachers have a clear vested interest in maintaining the literate/oral hierarchical opposition (viii, 84). The complicity of teachers in maintaining social relationships predicated on this opposition is complicated by disparities between how students of different social classes are taught. Jean Anyon (1980) was the first empirical researcher to track this disparity. Anyon's central hypothesis is that one's socioeconomic origin is a function of the quality of education one receives (see also Kozol 1991). The strategy for social reproduction can be seen not only in the way in which the working-class schools[4] fail to prepare students for any kind of postsecondary education but also in the way the pedagogies are shaped by what roles students from this social class are expected to play in their adult lives. Working-class schools train students to be workers—the students in these schools learn how to follow rules and how to memorize instructions; they learn that social reality is static and not of their construction. More concretely, they learn how to sound out, how to spell, how to identify the rules for commas. At the other end of the spectrum in what Anyon calls the "executive elite" schools, students learn how to govern. They learn how to challenge rules, how to shape them to the situation. They learn that reality is dynamic and that they are very much in charge of its construction—for instance, they negotiate assignments with their teachers. In writing, they do not concentrate on mechanics. They concentrate on the structure of what they write. By linking Bernstein's (1971), Brice Heath's (1983), and Lareau's (2003) research to the differences in kinds of schools, one can see that working-class students may do well in working-class schools, but they meet trouble in middle- or upper-class schools. From a different angle, they might do well in the lower grades but do

3. In contrast to unlegitimated intellectuals, from whom we rarely hear (see Bourdieu's [1984] discussion of autodidacts, 25-26, 99).

4. By working-class schools, Anyon means schools with a significant majority of students who have working-class mothers and fathers. Likewise, schools identified as middle-class, affluent professional, or executive elite have a corresponding majority of students from those social classes.

poorly as they advance through the upper grades (see Rose 1989). Brice Heath describes Nellie, a working-class black student, who was incapable of competing with her peers in her home community, where verbal skills and imagination were rewarded, but she "blossomed" in the primary grades, where the focus was on rote-learning. In sixth grade, Brice Heath reported, "she is floundering . . . but her quiet, polite way has thus far carried her through without a serious failure" (360). In light of Anyon's research, the conjunction of social reproduction and resistance theories take on additional complications because the kinds of reproduction and resistance will be different according to the kind of school—and according to the kind of students within each school.

In sum, the function of education is complicated. The levels of interpretation of the function go deep and their paths are full of switchbacks. One starts by assuming that the function is to impart knowledge and skills, but after a few years of teaching, one realizes that another function is to socialize students *through* various kinds of instruction by teachers who have been accordingly socialized. The socialization function itself has sublevels: teachers are trying to socialize students into behavior appropriate for further participation in the educational system, but on a more abstract level, teachers are trying to socialize young people into behavior appropriate for their participation in the adult world. The initiatory function is differentiated according to the student's gender, race, and social class differences—to mark the most obvious criteria. Women are taught to act differently than men, and working-class children are taught to act differently than the children of executive elites. The addition of race and sexual orientation to this formula (black, working-class, lesbian women) quadruples the possibilities of differentiation.

The final complication in reproduction and resistance theories concerns intention. To some extent, ascribing intention to a social system is metaphorical because social systems consist of groups of individuals, with each person belonging simultaneously to a multiplicity of overlapping and conflicting groups. In addition, the people and groups are dynamic, ever-changing over time. So a group is not a closed circle capable of a collective intent. That being said, there is in any social system a dominant group who direct not only national but also educational policy. One might note, for instance, that in all presidential cabinets—who effectively direct national policy—that at least 60% of the members have come from the elites (the top 1 to 2% of the total population). The figure averages 60% for the Democratic cabinets for the last 100 years and 71% for the Republican (Kerbo 1996, 174). David Smith has documented the corresponding influence of the corporate and elite classes on universities—the corporate and elite classes dominate the trustees and directly influence what gets researched and taught through their donations (as cited in Kerbo, 384). At some level, these people act as a group with intention. At my former university, there was a conscious intent to create graduates who would fill the gap between supply and demand of technologically sophisticated workers—in my field, of technologically sophisticated writers.

By contrast, the intent to teach working-class children the kinds of behaviors that will suit working-class jobs seems relatively unintended—or at least the strategy of presorting them through offering different kinds of instruction seems unintended. To appreciate the full dynamic of levels of intention in educational institutions, one simply has to imagine the social dance between parents, boards of education, administrations, support services, teachers, and students, all with different persuasive or coercive powers. Each group with its subgroups is capable of resisting to varying degrees the agendas of the dominant groups. These moments of resistance can turn against the resisting groups, acting against them as well as against the social reproduction agenda. Finally, the moments of resistance can serendipitously play into the hands of the dominant groups whose interests may not coincide with the interest of the overall social group.

I have sketched here only the most obvious factors that work with and against each other in the educational industry. Even these obvious factors create a kind of motivational jungle that would be daunting if we had not had quite a few people who have hacked out paths before us. Most of the work that has been done so far has focused on the structures and functions of the educational system as a whole (e.g., Bourdieu and Passeron 1990; Freire[1970] 1995; Katz 1975; Kozol 1991) or, within English studies on literacy in general (e.g., Brice Heath 1983; Gee 1996; Finn 1999; Stuckey 1991). In this book, I will be investigating the same subject but with a focus on writing instruction—as it is situated within English studies. I am particularly interested in how progressive writing teachers like myself maintain social class structures while thinking we are working against them.

ORGANIZATION

I begin with a discussion of social class and its function in writing instruction. Although the relationship is seen most clearly in the differences between working- and middle-class English, Lynn Bloom (1996) has argued in "Freshman English as a Middle-class Enterprise" that language use is only the surface signifier of differences between the working- and middle-class ethos. The real difference is deeper than language: it lies in the way of being, which Pierre Bourdieu (1984) describes as habitus, James Paul Gee (1996) as Discourse, and Basil Bernstein (1971) as social codes. In the rest of the book, I work primarily from the perspectives of Bourdieu, Gee, and Bernstein, who argue that these deeper differences contribute to the systemic failure of working-class students in a dominantly middle-class educational project.

I contextualize my critique of critical pedagogy within the common linkage between critical thinking and good writing. I trace the evolution of critical thinking into the cognitive and social strands of critical thinking and writing. Teachers in the cognitive strand focus on argumentation as the exclusive vehicle of critical thought. I argue (ironically) that the link between critical thinking and argumentation is a prototypical middle-class concept, marginalizing working-class students, whose habitus is in conflict with the assumptions of this strand. Consequently, writing courses devoted to what is basically argumentation for its own sake function subversively as a sorting mechanism based on privileging extrinsic knowledge (Bourdieu 1984, 23), that exists outside the area of instruction and is thus a property of the cultural knowledge and the linguistic and cognitive habits students bring with them.

As in the cognitive strand, teachers in the social strand of critical thinking privilege argumentation but for the larger purpose of promoting social justice. The critical thinking within this strand is not a function of informal logic and language; rather, it applies to a way of reading culture, of demystifying or de-naturalizing socializing narratives. It inherits from the cognitivists the primary assumptions of argumentation that conflict with the working-class habitus. In addition, the values and subjects of investigation (usually themes associated with identity and difference) tend to alienate working-class students for reasons that postmodern, middle-class teachers may dismiss as symptomatic of submersion in hegemonic discourse. A classic example would be the interpretation of working-class solidarity as conformity.

I devote the larger portion of my critique to the social strand of critical thinking and writing for two reasons: first, it is arguably the dominant mode of instruction for required writing classes; and second, it is as much a self-critique as it is a critique of how others teach. I am not dismissing the social strand—in fact, I consider myself up to my neck in it. But I am trying to understand some contradictions in the critical pedagogy agenda that work against it. The most startling contradiction is that the working-class students disproportionately resist and concomitantly have the fewest resources to circumvent or negotiate a critical pedagogy agenda.

As several scholars have noted (see a summary in Hardin 2001), accounts of student resistance are liberally distributed in the literature on critical teaching. In my research, I also read accounts of critical teaching that that did not slip into some form of the conversion narrative. These were teachers who listened carefully to their students, who investigated their students' worlds and respected their knowledge, who did not assume their students had been duped by the hegemonic discourse, who did not take it upon themselves to correct their students' materialistic desires. But these critical teachers were not the norm in the literature among teachers who linked teaching writing with teaching critical thinking (see Ellsworth 1989; Fox 2002). As teachers, we have been taught through the instrument of education that our degrees distinguish—to use Bourdieu's (1984) notion—our superior (critical) thinking. It's extremely difficult, if not impossible, to escape the socializing influence of the culture that surrounds us. As may become embarrassingly clear in the pages that follow, I certainly haven't done so.

Escape from the patronizing attitude naturalized through the educational industry is, I argue, a prerequisite for an authentic Freirean pedagogy. But far too many critical teachers have difficulty stepping outside the assumption that they know and their students don't. As well as having been socialized within the educational industry, critical teachers have the additional burden of their social justice mission which calls them to make of their classrooms, as Victor Villanueva (1993) put it, "an ideal site in which to affect [sic] change" (21). Of the well-known critical teachers who seem to have escaped their training and the patronage that comes from having a mission, Ira Shor's (1992; 1996) accounts of his classroom exchanges reflect a Freirean humility that makes room for students' knowledge. I was struck by Lori Robinson's (1993) reflection on her teaching in "'This Could Have Been Me': Composition and the Implications of Cultural Perspective," Carol Faulkner's (1998) respect for her working-class students and their reasons for enduring the gauntlet of middle-class education, Nancy Mack's (2006) imaginative assignments drawing on her students' working-class backgrounds, and by

Donna LeCourt's (2006) acknowledgement of her students' abilities to occupy multiple and shifting social class positions created through textual representations and social relations.

These teachers manage to negotiate the thin corridor between their social justice agendas and their students' purposes and home Discourses, no matter what their social class origins. I am not writing this book for teachers like Robinson, Faulkner, Mack, and LeCourt, teachers who have been alert to the perils of politicizing required writing classes. Practiced teachers like these might read this book and wonder about its purpose because so many, like James Zebroski (1992), have already warned against the dangers inherent in allowing one's political agenda to override one's primary responsibility to help students improve their writing, but the fact that the call for transformational pedagogy still dominates our professional literature (Fulkerson 2005, 60) suggests that the degree to which one should allow one's politics to leak into one's classroom is far from settled; in addition, I think we need to be concerned not about rhetoric and composition scholars but about the graduate students and instructors who do the bulk of our teaching. They do not have the privilege of release time for research and writing or the funds to attend our conferences and may consequently uncritically take up Villanueva's call and inadvertently transform the writing classroom into an ideological war zone in which working-class students are inevitably the losers.

APOLOGY

With one exception, any teacher I have mentioned in this book, I admire. (The exception is a graduate student who was drowning in his or her mission.) Several are my friends. In some way or another, we share the same project. What I am calling their mistakes have been, and may still be, mine. I have worried in writing this book about opening up their classroom dynamics for critique. At times, I have thought, no, don't do that. Sweep it under the rug. Don't write something from which your friends may take offense. I am edging into the territory where the private borders the public—at what point do I as a writer make my thoughts public and risk exposing not only the person I am writing about but also me as a member of the rhetoric and composition community?

The conundrum of this situation is that, as my very good friend, Bill Thelin (2000), has courageously argued, we learn from those moments that are dangerous. We don't learn when we run around patting each other on the back. Most frequently, when I found myself wondering, should I really say that?, I have written it. In a way, my remarks in this book are the suggestions I would make to a writer who had just read his or her paper aloud to a peer response group. Good peer response works when the critic is able to bring his or her criticism home.

2
SOCIAL CLASS

Before taking up the question of the intersections among social class, critical thinking, and writing instruction, I will analyze some of the problems of referring to social class in the Unites States. The major issues are the un-naming of class, its empirical status, what markers we use to distinguish the different classes, and what we call them.

In rhetoric and composition, we have the additional problem of abstracting class from the larger field of marginalizing status markers, the most common ones being race, ethnicity, gender, and sexual orientation. To imagine that any of these status markers operate independent of the others is naïve. To debate which marginalizing status marker deserves primary status is equally naïve. It temps the verbal equivalent of what Freire ([1970] 1995) calls "lateral violence," peasants fighting against each other instead of against the landowners (48). The social structure that feeds off marginalized social groups is our problem.

But in order to study any phenomenon, we abstract it from a larger field—or, said another way: we focus on the baseball heading toward us if we want to hit it. This is what I will be doing in order to analyze the intersections between social class, critical thinking, and writing instruction. I do not, however, imagine that these intersections can be understood fully if one forgets that one *has* abstracted the issue of class from the larger field of social marginalization. In practice, the analysis of class has to be reintegrated with studies of other methods of marginalizing social groups.

In my analysis, I will try not to repeat what may be common knowledge among scholars who have included in their study social class dynamics. However, I will carefully describe my own understanding of social class so that readers who may not have followed research on stratification theory will not misinterpret what I mean when I refer to different social groups.

ABSENCE OF CLASS

Writers commonly note the aversion in the United States to confronting issues of social class (hooks 1995; Scott 2009; Shepard 1998; Stuckey

1991; Tate 1998; Tate, McMillan, & Woodworth 1997; Villanueva 1998). Alan Shepard compares this silence to the frankness with which class in British universities is not only discussed but objectified through academic practices such as marking degrees with class designations and allowing only certain "classes" of people to walk on lawns assigned to their social class. The mythology and historical development of the United States are usually cited as contributing factors to our disinclination to focus on social class. The capitalist mythology works best when people can imagine that everyone has a reasonably equal opportunity to achieve success, which is considered the just reward of diligence, industry, and intelligence. Wedded to this bootstrap mythology is our pioneer heritage sanctifying Daniel Boone individuality and belittling group behavior. Consequently, Americans are socialized into thinking that only in other countries (like England) are citizens pinned by class to the wall.

This silence in our field on class issues has been reflected by the number of sessions in our CCCC and MLA conventions. MLA seems to have been particularly silent on the issue. Shor (personal communication) reports that he co-chaired some sessions in 1971 and 1972 with people like Fredrick Jameson, Norman Rudich, Paul Lauter, Richard Ohmann, Louis Kampf, and Richard Wasson in which class was present as a subject but it wasn't named in the session titles. Deborah Holdstein (personal communication) said she was in a session on the literature of the War in Vietnam in the late 70's in which class became part of the discussion but wasn't mentioned in the title. Renny Christopher (personal communication), who researched this issue in order to initiate a discussion on class at MLA, reports that from the years 1993 to 1996, only three sessions were devoted to class—one in 1993, one in 1995, and one in 1996. In her article, "Freshman Composition as a Middle-class Enterprise," Lynn Bloom (1996) wrote that after issuing a call in 1993 for papers on race, class, and gender in composition studies, she received one proposal on class, compared to a dozen on race and 94 on gender. She noted that "the C-word" rarely was named in paper titles until 1994 (657). Lest we imagine that class as a subject has escaped the closet in the current century, Cindy Selfe in 2009 (personal communication) told me that in their responses to surveys, people volunteering to offer their stories of literacy to the Digital Archive of Literacy Narratives consistently left blank the question asking them to identify their social class, in contrast to other social category markers such as race and gender.

I have conflicting responses to these claims and evidence of a cultural aversion to class issues. My history as a working-class academic has made me suspect these claims of silence. I personally enjoy reading about, thinking about, writing about, and talking about class. To a certain extent, some of my friends and colleagues seem interested in the discussion. On the other hand, I have been in many conference sessions where others report difficulties in bringing up these issues with their colleagues and students. The brief research that I have done and reported above suggests a certain silencing, albeit one I have not experienced in my professional life.

On the surface level, whether one avoids or welcomes the discussion might be a function of one's social class origins. Lawrence MacKenzie (1998) writes that "an oppressive requirement of being or appearing middle class is to avoid social class talk . . . to speak of class is itself, conveniently, déclassé" (105). MacKenzie is a working-class academic labeling a feature of middle-class membership. When you come from one class, it's tempting to notice negative features of the "other" class—MacKenzie and I probably interpret the avoidance of class-talk as a negative feature in the middle-class ethos. Nevertheless, there might be something to MacKenzie's claim. If you belong to the middle or upper classes, you might prefer not to mention the social game that has made it easier for you to be where you are. If you are a middle-class academic with lower working-class origins, you might be pleased to write and talk about class as an oblique self-congratulatory gesture for having "risen."

Working-class people are on the obverse side of the true-grit coin. Class recognition is deeply embedded in their identities particularly among those who have crossed from one social class into another (see Giroux 1991). They understand that class is very much a part of the reason they are poor. Laurel Johnson Black (1995) writes of the division between the classes: "I learned that the stupid rich bastards always underestimated us, always thought we were as dumb as we were poor, always mistook our silence for ignorance, our shabby clothes and rusted cars for lack of ambition or enterprise" (15-16). That's recognition of class.

I conducted some casual research in 1999 that supports this hypothesis on the relationship between social class origins and one's willingness to discuss the issue. As a consequence of another discussion on WPA-L (the Writing Program Administrators listserv), I invited members to send me twenty- to thirty-minute reflections on their social class origins and what their origins had to do with how they teach. Within three weeks

I received twenty-five replies. Of the twenty-five responses that I received, thirteen of them had been written by academics with working-class backgrounds, eleven by academics with middle-class backgrounds, and one by an academic who was a clear border case. That is, 56% of the responses were from working-class academics. Other research that I have done on the relationship between social class origin and academic position suggests that 34% of rhetoric and composition professors have working-class origins, 66% have middle-class origins, and fewer than 0.5% have upper-class origins (N=291). These statistics imply that working-class are more willing than middle-class academics to put class on the table.

There are, however, further complications for working-class academics. To put the case bluntly: becoming an academic is our way of escaping the working-class.[1] In order to escape, we have had to make over our identities in serious ways. I suspect that many of us would prefer not to make apparent what we were trying to cover (see Tate 1998; Sullivan 1998). James Paul Gee's (1996) theory of primary and secondary Discourses provides a way of understanding the kind of identity revision I am suggesting. In Gee's usage, a Discourse means more than one's linguistic habits. It is similar to Bourdieu's concept of habitus. As well as speech codes, it includes things like dress codes, grooming habits, ways of moving and eating, housing, furniture, music, movies, reading material, sports, etc.—in short, everything that contributes to one's identity as a member of a social group. The primary Discourse is the discourse of one's family and close friends who share one's habitus. The secondary is the public Discourse—the one we "put on."

Working-class academics learn to submerge their working-class or primary Discourse as they struggle to replace it with the adopted middle-class Discourse. Pat Sullivan (1998) writes about this covering over and coming out:

> I had never told either friend nor anyone else I have ever worked or studied with in academe for that matter, the social and familial circumstances of my life. In trying now to take measure of that silence—now, as I am about to break it—I am stalled by a wave of fear. There is a voice inside me that has been there as long as I can remember, and it says that it isn't right to speak

1. This sweeping generalization embeds three major sub-claims that I will not argue but I want to make explicit. They are 1, that my motivation has been similar to the motivation of other working-class academics; 2, that being an academic is one way of escaping one's working class condition; and 3, that the working-class condition is something one would want to escape.

> of private matters, family matters, money matters. That there will be hell to pay. That it is safer to "pass," to keep projecting a self who has made it, than to disclose the conditions of getting there. (232)

Once we have made it, once we have learned to call dinner lunch and supper dinner, to chew with the front of our mouths (Bourdieu 1984, 191), and what to do with the different glasses and extra dinnerware on the table, we do not like to be called back to our social class origins. We have worked hard to erase them. Even now, when I am sixty-five and dominantly middle-class, I watch others to find out how to eat when I go out to expensive restaurants. There are certain things I can't seem to get "right."

Gary Tate (1998) has sharply brought this conflict into focus in "Halfway Back Home." He describes in this searing essay his struggle to disguise his social class origins.

> My rituals of denial continued through graduate school and into my teaching career where they assumed many forms. I have tended until very recently not to sympathize with students who have trouble overcoming the odds of a deck stacked against them . . . . I have also tried to prove that I belonged by adopting the highest possible standards. The better the college, the higher my standards—read "degree of insecurity." During one two-year period at a good liberal arts college, I taught 179 students. In that two-year period, I gave two A's. . . . I lectured on the latest composition theories while penalizing first-year students for the slightest deviations from Edited American English. . . . The scorn I felt for my past, my parents, my students was nothing compared to the scorn I felt for myself, which explains, in part at least, my final act of denial: a turn to drugs. (254)

One can read deeply into Tate's brief story here. The desire to look middle-class might make one more insistently middle-class than those who are born to it. Freire ([1970] 1995) comments, "It is a rare peasant who, once 'promoted' to overseer, does not become more of a tyrant towards his former comrades than the owner himself . . . . In order to make sure of his job, [he] must be as tough as the owner—and more so" (28). I underscore Tate's *rigorous* practice with Freire's analysis of the overseer to mark the strategies that working-class academics who want to erase their primary Discourses employ. By silencing the working-class students who are also trying to escape, they are silencing their working-class selves—the ones lying underneath the veneer of working-class academics. Tate is right: drugs kill the pain and keep us quiet. Of course middle-class drugs come in various disguises.

This reading of working-class academics contradicts my speculation that working-class academics seem more open than middle-class academics to discussions of class. This contradiction is perhaps accounted for by the development of working-class studies. Although teachers like Ira Shor, Barbara Foley, Richard Ohmann, and Paul Lautier were writing books and articles and giving papers on class issues from the 70s on, class didn't become a named focus like black studies or women's studies until the 90s. A book edited by Jake Ryan and Charles Sackrey, *Strangers in Paradise: Academics from the Working-class* (1984) seems to have been the cornerstone on which this new focus was built. *Strangers in Paradise* was a collection of essays in which academics from the working-class, one, admitted, that they did indeed have working-class origins, and two, vented the anger they felt over having denied their origins. Perhaps within this anger lay the suspicion that everyone from the higher classes (that is, the significant majority of their colleagues) saw beneath their pretense. Bourdieu (1984) calls pretenders like the writers in *Strangers* "parvenus" or "late arrivals" (164); Gee (2004) first called them "latecomers" but later switched the term to "authentic beginners"—distinguishing them from "false beginners" who were born to the game. The parvenus have to learn the rules from scratch. Bourdieu's point is that those who were born to the condition *always* recognize the parvenus, who reveal themselves, like Tate, by over-monitoring their behavior.

No matter how hard working-class academics struggle to disguise their origins, the false-beginners know. To explain this kind of paralinguistic perception, Gee (1996) refers to studies that explain how "real Indians" know each other. It's not in what one says or looks like; it's a way of being. One real Indian recognizes another as well as the pretenders, who, no matter how hard they try, will always be outsiders to Indianess (129-130). I understand this: I can recognize working-class academics without hearing them speak. I recognize the middle- and the sprinkling of upper-class academics by the law of exclusion—and I assume that's how they recognize us. This fear, then, of being *seen through* might account for the anger expressed in *Strangers in Paradise.* As Pat Sullivan has made clear, it also accounts for the silence.

But in the early 1990s, conditions changed. Janet Zandy (1990), one of the important forces to bring about this change, helped to give voice to working-class women by putting together *Calling Home: Working-class Women's Writings.* In 1994, she collected essays from working-class writers and published them in *Liberating Memory: Our Work and Our Working-class*

Consciousness. The writers in both books are given space within which to recall the experience of growing up working-class. Continuing the tradition which began with *Strangers in Paradise,* C. L. Barney Dews and Carolyn Law (1995) collected essays from working-class academics and published them in *This Fine Place So Far From Home: Voices of Academics from the Working-class.* In 1996, John Russo and Sherry Linkon developed the Center for Working-class Studies in Youngstown, Ohio and have been sponsoring biennial conferences on working-class studies. During these years (1995 and 1996) sessions devoted to class and working-class studies began to appear at CCCC, most of which were organized by Ira Shor and his graduate students, Caroline Pari, Leo Parascondola, and Eileen Ferretti. In 1995, Doug Paterson, Mary Machietto, and I organized the first Pedagogy and Theatre of the Oppressed conference in Omaha—which became a yearly event and a place for artists, teachers, and cultural workers to meet.[2] In 2004, Linkon, Benjamin Lanier-Nabors, and I edited a special edition of articles on social class in writing for *College English.* The cumulative effect of this activity has been a recognition that those of us who were ear 'oles, who worked to make ourselves over, could freely acknowledge our working-class origins. This liberation of our memories is again captured by Gary Tate's (1998) description of his own liberation when he stumbled across a revised version of *Strangers in Paradise.* He writes

> To put it simply, I found myself in that book. As I read account after account of the pain and discomfort and anger of these authors as they had tried to negotiate the rituals and traditions of the academy, it was as if my entire previous life changed in front of me. "Yes!" "Yes!" I kept saying to myself. I am here, in this book, on almost every page. (255)

In short, this legitimation of our working-class identities has spread, accounting, perhaps, for the higher proportion of working-class academics who responded to my call for reflective narratives on their social class origins and how they teach. The subject of class may still be déclassé for many academics, but not for a rising proportion of working-class academics. In the latest turn of the naming game, it is now cool to be working-class (well, to have come from the working-class and maybe dress working-class—but not *be* working-class). It's enough to make me buy a suit.

2. Doug and Mary were really the leaders of this project, which focused more on theater and the work of Augusto Boal than on writing and Paulo Freire.

ONTOLOGY OF CLASS

Lower, working, lower middle, middle, upper middle, upper, underclass, working poor, homeless, overclass, subordinate class, dominate class, elites, non-elites, dominant elites, professional, intellectual, professional-managerial, proletariat, petite and petty bourgeoisie, bourgeoisie, exploited, exploiters, capitalist, fast capitalist, corporate, bureaucrats, middle management, white collar, technocrats, blue collar, labor, manual, higher non-manual, lower non-manual, skilled manual, unskilled manual, peasants, landowners, propertied class. Class matters, but what is it?

I tend to be casual about which labels I use, because class is a conception. Class does not have a concrete reality that is "out there" somewhere to be recognized, analyzed, and described. But a good deal is at stake when one names the different classes. Dividing people into lower and upper-classes assigns moral and personal values to members of either group. This value reverberates throughout our language, chiefly through the associations that lie within the words. George Lakoff (1987) in *Women, Fire, and Dangerous Things* has analyzed how metaphors like lower and upper are embodied—they come *out* of our bodies (262). In this case the feet are the lower; the head is the upper. The feet are for walking; the head is for thinking. The lower class is for doing; the upper-class is for organizing, directing. The lower part of the body is for sexual reproduction and excretion; the upper part of the body is for circulating the blood, getting air, seeing, smelling, tasting, hearing, thinking. The lower class is for having babies and taking care of various waste products; the upper-class eats and thinks. The historic association and ideological power of the lower/upper dichotomy is instantiated through the visually concrete descriptions of hell and heaven. It is of course no accident that the *head* of heaven is King and male (and white). One could go on. It doesn't take a lot of thinking to understand what kind of conceptions are naturalized by speaking of the lower and upper classes. No matter what words one chooses, one will load into those words a series of associations and a babble of previous voices that have used them in specific contexts (Bakhtin 1981). So a lot is at stake when we name class.

Likewise, much is at stake when we choose criteria by which to identify the different classes. The basic division seems to be between the Marxists and everyone else. The Marxists in turn are divided between the big Ms and little ms with the latter sliding into everyone else. The classical Marxists insist that class has to be defined (following Marx's analysis of the capitalist social structure) by a group's relationship to the mode of production. Marx divided these into the bourgeoisie and proletariat—or those who own the means of production and those who sell their labor for less than it is worth and consequently create profit for the bourgeoisie. Several quasi-classes slide into or out of these two major classes—e.g., state bureaucrats, middle managers, and so on. The problem of the in-between classes has grown more complicated with what has become known as fast capitalism within which well-paid laborers like computer programmers may have significant investments in the companies they work for.

The "everyone else" group includes theorists who have looked outside the mode of production for ways of defining social classes and their relationships. Bourdieu (1984) broadens the notion of capital (and who owns it) to include such things as educational, social, cultural, linguistic, and symbolic capital. As well as opening up a multi-dimensional perspective on one's social position, Bourdieu theorizes a dynamic relationship between the different kinds of capital; thus, one can trade economic capital for educational or symbolic capital. Because of this dynamic relationship and multi-dimensionality, Bourdieu thinks of social class as situational, relational, and spacial. And finally, Bourdieu doesn't speak so much of a social space as of a trajectory, one's path from birth to death.

Another element has to be taken into account when we speak of social class—the position from which we speak. On a broad level, this might refer to physical location. As I noted above, discussions about class take on a different tone if one is speaking from within England or the United States. Class is an accepted, unproblematic notion in England; in the United States, it is sometimes fiercely denied as a useful concept. American theorists have to work their way through an ideolology that attempts to erase class. A case in point might be found in Richard Ohmann's (1982) challenge to Basil Bernstein's theory of linguistic codes characteristic of different classes. Ohmann faults Bernstein's analysis because Bernstein fails to account for the class categories that he uses—that is, he fails to establish and justify his criteria and the categories. Rather, Bernstein, according to Ohmann, blithely assumes the categories of working and professional-managerial classes and on the basis of his research assigns codes characteristic of these different classes. Bernstein does seem to have taken these social classes for granted, but part of the difference between Bernstein and Ohmann might be accounted for by the difference between England and the

United States. Bernstein makes assumptions that Ohmann insists must be accounted for.

Self-referentiality is also at work when one tries to map the position from which one speaks and thinks about class. A person who comes from the upper-classes simply can't understand class the way I do. And I can't understand class the way someone who comes from the welfare class does. We can communicate, but our conceptions of class will never be the same any more than a flatlander's idea of a hill will be like a mountaineer's.

OBJECTIVITY OF CLASS

In our postmodern era, it seems unproblematic to claim that class is only a conception, an "in here," not an "out there," but "class is a conception" are fighting words if one is (as I was some years ago) at a conference on working-class studies. One might as well say in a conference devoted to African American literature that race doesn't exist, that it is a conception and exists only as a social condition in the speaker's mind. I need to emphasize: calling class a conception doesn't mean it doesn't exist; it simply doesn't exist "out there." Strictly speaking, nothing (no-thing) exists "out there." This line of thinking stretches from the sophists (Enos 1976) to the hypermodernists. Naming-words point to categories of things, ideas, or actions. Any category is by definition not "out there." A name referring to two similar items refers to a mental concept of those two items with the concept being built from features the items have in common while ignoring their differences. As names for categories point to groups with larger membership, the name increasingly points to something that is increasingly "in here" rather than "out there," that is, one is describing more and more one's way of linguistically organizing reality rather than to organizations that have an objective existence (see Lakoff 1987; Moffett 1968; Rosch 1978). One's way of linguistically organizing reality is of course determined by one's social existence, which includes the language into which one is born.

This theory of classification is in opposition to an Aristotelian notion of classification—which Lakoff (1987) calls a folk theory of classification (5). Aristotle imagined that there were real categories of things in material reality—as did Rosch (1978), until she began looking closely. Aristotle felt that through a careful and disinterested investigation of the examples of things, one could identify and describe the natural groupings—or categories. But as postmodernism has insistently claimed, what one sees depends on who, from where, and why one is looking.

In contrast to the Aristotelian conception of classification, contemporary theorists work from some variation of prototype theory, developed by Rosch (1978) and meshing with Wittgenstein's theory of "family resemblance" (1953, Aphorism 67). In essence, names are naturalized when they point at instances that fit a prototype of a category, like pointing at a robin as an instance of bird. But when names point at instances near the edges, like toward an emu, conversation gets messy—which may explain Ohmann's critique of Bernstein's notion of class.

I have explored classification theory to explain what I mean by class. I am referring to a system of social relationships within which people act toward each other as if the groups *did* exist—as in their minds, they do. There is certainly some kind of objective hierarchy of social positions based on a differential distribution of resources and constraints that govern any individual's choice of actions (Breen and Rottman 1995). I am imagining this hierarchy as existing on a continuum from few resources and maximum constraints (e.g. prisoners) to maximum resources and few constraints (e.g., elites), but as Bourdieu (1984) makes clear, a linear model is an abstraction/distortion of a more comprehensive model that maps an individual's social space on the basis of several kinds of capital.

When I am referring to class, I am primarily sorting on the basis of a person's occupation, level of authority, assets, level of education, and social relationships. I do not imagine that these are the only markers of class, but they are common ones people recognize. Following Bourdieu's distinctions, adopted from Marx, I will frequently refer to class fractions. The most useful class fractions are the economic and intellectual fractions that exist in opposition to each other within any particular class. Adopting class distinctions used by Kerbo (1996, 12-13), I will generally break working and middle classes into lower, middle, and upper within each larger grouping and to the upper class as a class by itself. I will also refer to the corporate class—the upper class that has to work.

A few caveats are in order. First, the divisions are affected by the environment: class systems are different in urban, suburban, town, and rural areas. In general, one could say that in denser populations, the degree of difference is higher and the segregation is more complete—class is ghettoized. Second, farmers and ranchers cross class lines. Their class depends on the size of the farm or ranch. They are all self-employed.

They are in charge of their own time and work. Increasingly, more of them have college degrees. Although they engage in manual labor, assigning them the same class positions as non-rural workers does violence to the farmer's or rancher's habitus. And third, references to education is relative to the times. An M.A. in 1950 would be equivalent to a Ph.D. today. An M.A. now would equal a B.A. in 1950. Bourdieu (1984) describes this process of shifting values in terms of the privileged classes continuously having to raise the bar in order to maintain their "distinction," or their distance from the common people. He calls it the hystereisis effect, something like keeping a carrot suspended by a pole on string in front of a mule (142).

Finally, this entire discussion might have been avoided if I had simply quoted Elspeth Stuckey's (1991) paraphrase of W. Lloyd Warner: "You know your class by who invites you to dinner" (4).

3

LANGUAGE, CLASS, AND CODES

Language is a particularly effective mechanism for maintaining distinctions among social classes because it functions both to communicate and signal identity, with one function frequently disguised as the other. Teachers, for example, may correct working-class students' deviations from the conventions of middle-class English, telling the students that the errors make their writing difficult to understand when in fact the teachers are correcting social class behavior manifested through language codes. Behind this masking lies the clear message that the social groups speaking through these "incorrect" language codes are incorrect social groups.

WORKING-CLASS ENGLISH

Although working-class English is one of the more insistently devalued dialects, it hasn't received much press because it's not interpreted as a dialect. Beyond Bernstein's (1971) research of working-class English at the level of codes, we have to date no systematic study of working-class English as a dialect. Although we are surrounded by it, working-class English hasn't been considered an object worthy of study.

But middle-class speakers recognize working-class English when they hear it. I have recorded snippets of language from working-class speakers in my family. They will say, for example, "you done good" for "you did well." We might guess that working-class speakers don't distinguish between adjectival and adverbial forms if the distinction carries no additional meaning other than social class membership. This rule would be a linguistic expression of Bourdieu's (1984) thesis that the working classes, because of their close relationship to necessity, privilege function over form. Within the dominant classes, where class membership is signaled by one's distance from necessity, the obverse is true. Form is privileged because through it, speakers signal their social class membership.

A working-class speaker will use "done" rather than the middle-class form, "did." It seems that when there is a spelling change between the present and past tenses (a strong verb), working-class speakers will substitute the past participle (done) for the past tense (did). My father, for instance, said the following when referring to the setting sun: "I seen her set," and this, when referring to the temperature: "It was fifty-two last I seen." Similarly, he said, "two more of 'em [eagles] come from over there." Perhaps the working-class rule is this: if you're uncertain, use the past participle for both past and perfect tenses. When using it for the perfect tense, drop the auxiliary. In any case, use the most common form of the verb. People will understand what you mean.

My father uses "has got" for "has" ("That's got milk in it"), the pronoun "them" for both "them" and the demonstrative "those" ("Well, I hope them groups can make a go of it"). Working-class speakers freely use the verboten double negative ("Oh, he didn't have no money") and double verbs for emphasis ("I went and pulled him [a fish] in the net and the bottom opened and he got right out"). As many linguists have noted, the working-class "ain't" is a useful contraction for "am not," "is not," or "are not" ("He ain't going") or for "has not" and "have not" ("We ain't got no room for that"). "There is" works perfectly well for "there are" ("There's some nests [the eagles again] down along the road. She taking them down there inside someplace"). Notice that by saying "She taking," my father feels free to dispense with the auxiliary in the progressive tense. Phrases are perfectly functional ("Warm today"). And the object forms of personal pronouns work equally as well in the nominative case ("Me and Gordon used to do it like that").

In these and many other examples of working-class English, a middle-class interlocutor would clearly understand what the speaker meant. The middle-class speaker might, however, imagine that the working-class English is inferior because it is less precise, not having the more variable forms available for different shades of meaning. But the difference is largely one of codes that signal social class membership. For example, a few years ago I had a graduate student teaching assistant (I think she came from an upper-middle class background) give me a sample of her student's in-class writing and complain that she didn't know what she could do to help him because his writing was so bad. I asked what was wrong with the writing, and she said, "It's just incoherent." I have reproduced the first page in figure 1.

By lack of coherence, this frustrated GTA meant she couldn't understand the passage, so I read it aloud to her and asked what she didn't understand about it. She was of course nonplussed because the passage

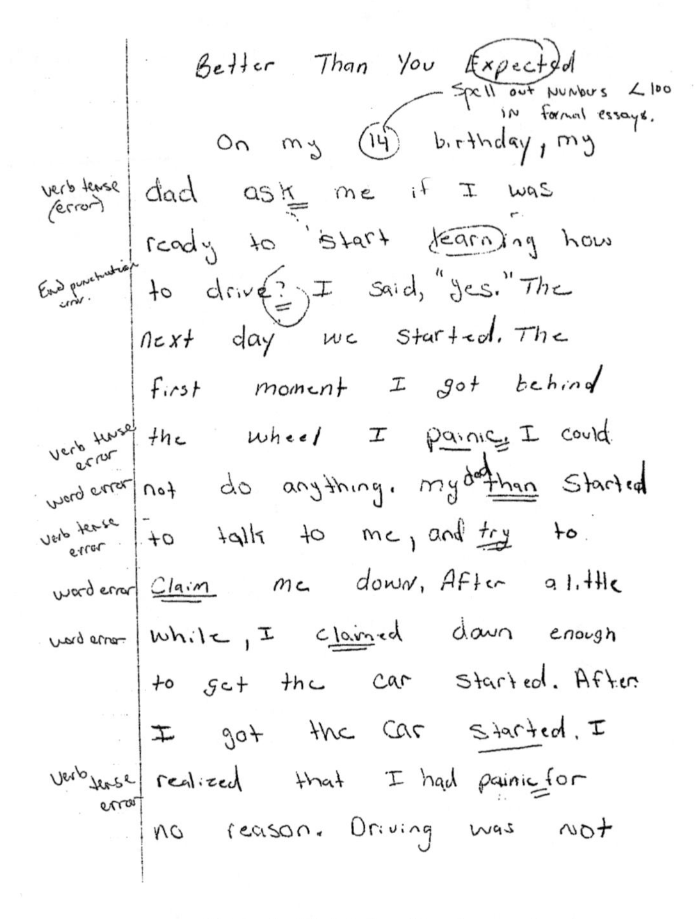
Better Than You Expected

On my 14 birthday, my dad ask me if I was ready to start learning how to drive? I said, "yes." The next day we started. The first moment I got behind the wheel I painic. I could not do anything. my dad than Started to talk to me, and try to Claim me down. After a little while, I claimed down enough to get the car started. After I got the car Started. I realized that I had painic for no reason. Driving was not

Figure 1. Mislabeling "incoherence."

was quite clear—as were the kinds of "errors" the student was making, the dominant one stemming from a lack of practice in translating speech of one social class into writing in another. The GTA was using the "incoherent" label to mean "incorrect," by which she really meant, "not from my social class." I had to struggle to help this GTA because she did not want to admit that she was confusing social with linguistic codes.

BERNSTEIN'S ANALYSIS OF CLASS CODES

Bernstein has come closest to a sustained analysis of class codes. His purpose was to contrast working-class with professional-managerial class English by emphasizing the relationship between language, cognition, and social class. By working-class, Bernstein meant semi-skilled and unskilled laborers with the man's education level at the grade school or possibly high school level and the woman's generally at the grade school level. Middle or professional-managerial level were classes above that. Although Bernstein's categories of class were constructed in Britain in the 60's, many writers have found in his analysis useful ways of understanding the differences between working- and middle-class languages and their relationships to school achievement (see Davies 1995).

The first and most enduring statements of his analysis were presented in *Class, Codes, and Control Vol. 1-3* (1971). The analysis was based on extensive tapes of discussion sessions and on research conducted by his associates at the Sociological Research Unit, University of London (*Pedagogy*). The key for Bernstein is the dialectic relationship between social environment, cognition, and language. These factors shape and are shaped by the others. Parents' work situations affect their thinking patterns, which in turn influence their language habits—for example, whether one learns how to arrange parts of a sentence hierarchically or serially; the degree to which one uses qualifiers; or the extent of one's vocabulary. Language habits shape how one sees the world, and one's vision of the world influences the social world in which one lives.

Bernstein bases his theory of class, codes, and social control on a series of dichotomies that I have organized into social, cognitive, and linguistic domains (see table 1).

Social	*Cognitive*	*Linguistic*
Communal/individual Closed/open role systems Positional/personal-oriented families	Content/structure of objects	Restricted/elaborated code Public/formal language Implicit/explicit Expressive/verbal

Table 1. Organization of Bernstein's dichotomies

The first set of Bernstein's dichotomies describes the environments in which children grow up. A communal organization means the family exists within a close network of family and friends; socially, they stay within that close network. The communal organization describes most working-class families. An individual-oriented organization extends relationships outside the immediate community of family and friends. Identities are individualized within a large network of social relationships that can cross not only neighborhoods, cities, and states but also countries. The individual organization describes the situation of most professional-managerial class families.

In a closed role system, individuals' identities are highly regulated within socially prescribed roles: i.e., the priest, a father, a mother, an aunt, a sister, a teacher, a police officer, a president. In an open role system, individuals negotiate their roles within the social system. The roles are as much self-constructed as they are socially constructed.

The role function aligns with positional- or personal-oriented families. In the former, children are taught to respond to the position an individual occupies (a father, an aunt, a boss); in the latter, children are taught to respond to who the person is. People are not granted authority by their position but by how they present themselves as people with persuasive powers.

These three modes of social control have a dialectical relationship with an individual's cognition. People governed by a communal, closed, positional social system tend to focus on a phenomenon as it exists existentially, to the *content* of the object, to its existential meaning. People governed by an individual, open, personal social system tend to see an object's meaning through re-*cognizing* its structure and comparing that structure to the structure of other objects. An object gains meaning through its similarities and differences with the structures of other objects. It also gains meaning by its relationships with other phenomena that constitute a structure at a higher level of abstraction. In sum, it gains meaning systemically rather than existentially.

These different cognitive orientations could be exemplified by comparing the work situation of line workers in a factory to the situation of CEOs of a large company. Factory workers are trained to focus on the specifics of their jobs. In a well-run assembly operation, workers don't have to know what is happening farther down the line. Understanding the relationship of the different operations is the manager's responsibility. Higher in the hierarchy, CEOs are the managers of managers. CEOs don't focus on details. They focus on the systems of the respective corporations. They have to know not only about different parts and how they relate to each other within the organizations but also how the organizations exist within the web of relationships between other corporations and the larger social system. The CEOs have to understand, for example, how to contribute to presidential races. For them, it is the larger web of existence that gives any part meaning. For the factory workers, a part has only its existential meaning. So does the closed community have only existential meaning—in the family, what counts *is* the family. So does the role—a foreman is a foreman, a teacher is a teacher, and that's that.

The social and cognitive orientations have a dialectic relationship with linguistic modes of social control. These modes are carried in restricted or elaborated codes. In comparison to an elaborated code, a restricted code offers few word choices, limited use of modifiers, and as we have seen above, simplified syntax. An elaborated code emphasizes word choice, careful modification, and a variety of syntactic structures to show relationships between sentence elements.

The degree of linguistic restriction or elaboration characterizes the difference between public and formal language. By public language, Bernstein means the kind of language that tends toward formulaic utterances, the significance of which is clear to members of a closed community. When these utterances become rituals, they have no individuated meaning. Interlocutors don't really pay attention to particular words or syntax. They know how to respond almost without thinking, so that both the utterance and response are ritual exchanges. "Hi, how are ya?" "Fine, how are you?" Heads nod. "Fine, thanks," and both speakers go their ways. This kind of discourse depends on the interlocutors knowing the ritualized meaning that lies behind, between, or around the words. The same could be said of a priest's ritual utterances: e.g., *Gloria Patri, et Filio, et Spiritui Sancto, Sicut erat in principio, et nunc, et semper, et in saecula saeculorum. Amen.* No one has to pay attention to what is said—as long as it is always the same thing in the appropriate rhetorical situation; in this case, someone is about to sprinkle someone else with water.

By formal language, Bernstein describes the kind of language in which the speaker or writer cannot depend on the audience's shared understanding of the meaning behind utterances. Consequently, the meaning has to be textually signaled in precise nouns carefully modified by adjectives, in precise verbs modified by adverbs, in relationships between the parts being signaled by subordinating conjunctions or syntactic embeddings. Bernstein calls this "formal" language because meaning is carried in the form rather than through a communal/public understanding. Thus for users of a formal language, if readers from a different social group, geographical area, or era read a text, they will be able to understand with some degree of accuracy the information the speaker or writer was trying to communicate. Legal documents are extreme examples—the purpose of the density of legal documents is to reduce ambiguity. We have learned from postmodern examinations of language that all utterances are open to alternative readings, but formal language struggles toward unambiguous meaning. So does academic discourse—particularly scientific academic discourse that depends on accurate information and repeatability of an experiment.

By the *implicit/explicit* dichotomy, Bernstein refers to the shared meaning that lies unstated in working-class language and the meaning that is made explicit by the language in middle-class language. Working-class discourse makes use of paralinguistic signals—gestures and tones—understood by the members of the close community. Bernstein calls this *expressive* language—in contrast to the rhetoric and composition community's interpretation of expressive language as emotionally saturated discourse. Ritualized and working-class discourse meet in the use of *expressive* language because the ritual utterances of a priest gain meaning through gestures and props rather than explicitly through verbal language. This congruence of ritualized and working-class discourse helps to explain the apparent contradiction of locating public language within formal acts like a Catholic communion in contrast to the formal language one expects in an academic journal.

These linguistic dichotomies act back on the social domain. Users of elaborated code make themselves explicit through language. People who do not know how to make their meaning (and significance) explicit through text do not count in the middle-class world. In working-class communities, individuals do not have to signal their meaning through text; the meaning is already known through the role into which they are born—for instance, as a man or a woman, as a laborer or as a housewife. One doesn't have to find a way to say who one is. One is already said.

OBJECTIONS TO BERNSTEIN

Bernstein's analysis of the relationships between social class, linguistic codes, and social control has met frequent criticism on the basis of his dichotomies, his notions of hard categories, the plasticity of the boundaries between categories, his obscurity, and the hierarchized values implied by his scheme (Walford 1995) in spite of Bernstein's (1996) insistence that he was attempting "to show that modalities are not simple dichotomies but oppositional forms, and that each has a range of realizations . . ." (4). By framing these as more than simple dichotomies, Bernstein emphasizes that the antinomies influence each other; for example, public language influences formal language which dialectically reshapes public language. The "range of realizations" positions

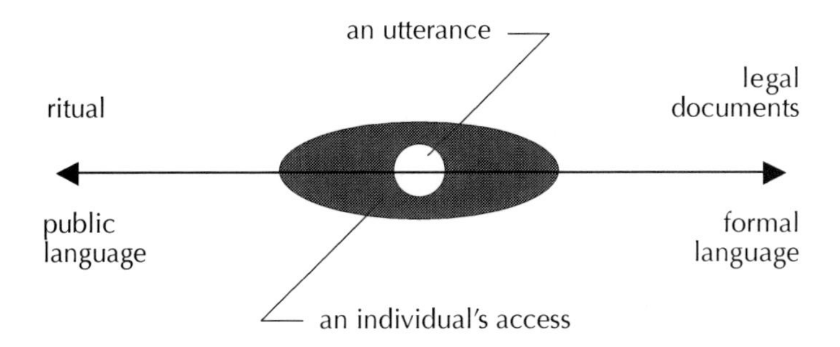

Figure 2. Scheme of the relationship between public and formal language

speakers somewhere within a dialectic: for example, whether one relies on public or formal language is not an either/or situation—one is situated somewhere on a continuum within an extreme form of public language (e.g., ritual utterances) lying on one end and an extreme form of formal language (e.g., a legal document) on the other. Individuals have access to different degrees of public or formal language, and they adapt their particular utterances to fit specific situations. Finally, any extended utterance may have within it different aspects of public and formal language (see figure 2).

Rather than think in terms of hard categories and rigid boundaries that seem to overdetermine both language and speakers, it is more useful to frame Bernstein's analysis within Bakhtin's (1981; 1986) theory of voice and heteroglossia—that in any given utterance, a multiplicity of voices can be heard. So it is with speakers.

The objection to Bernstein's theory that has gained most currency in the United States was raised by Richard Ohmann (1982; see also Bourdieu 1991, 53). Ohmann's objection should be interpreted within the context of a contradiction that has traditionally complicated theories of the left. The problem is this: to move toward an egalitarian social structure, the working classes have to understand how social systems work—and more particularly, the structures that have created the system of oppression. The working classes have to understand what Freire ([1970] 1995) calls the anthropological nature of culture, the fact that culture is made by people—generally by those who are in the dominant classes. Further, the working classes have to learn that through (and only through) collective action they can transform existing oppressive conditions into more egalitarian social structures. Oppressive social structures, however, create social conditions such that the oppressed will not have the linguistic and cognitive resources necessary to understand and transform their conditions of existence. This knotty contradiction leads to the danger of vanguardism or Leninism—a claim that the only way to transform existing conditions is to have intellectuals from the dominant classes show the dominated classes how to emerge as a collectivity, overthrow the oppressors, and put in place a non-oppressive social structure. The contradictions within this plot are obvious and lead to messy situations.

Ohmann (1982) argues that Bernstein's theory leads to a negativist philosophy because left to themselves, the working-classes will not have access to elaborated codes. Unless enlightened by the intelligentsia (whose motives must be questioned), the codes of the working-class will always be restricted and consequently, so will their thinking. We are left with a deficit/competence dichotomy (Bernstein [1996, 153] calls it a deficit/difference)—that the only way to empower the working-classes is to teach them the codes of the dominant classes. Thus, the codes of the dominant classes are the desired ones and the codes of the working-classes are the deficient ones. They have to learn middle-class codes because working-class codes simply won't cut the mustard.

The deficit interpretation of working-class codes lies at the base of John Trimbur's (1993a) objection to Mike Rose's *Lives on the Boundary* (1989) and of J. Elspeth Stuckey's (1991) objection to just about everybody. Leftist scholars struggle to dissolve the hierarchical oppositions and leave in place something like the ideal anthropologist's way of understanding different cultures—that is, no culture is superior to any other. Different cultures are simply different. Likewise with codes. To hierachize them on a deficit/competence continuum is to impose on the subject the position from which the evaluator is interpreting it.

In spite of my sympathy with this critique, I think that the attempt to equalize all discourses leaves one hanging. The overarching structure of competing discourses complicates the ranking. One simply can't ignore that the discourses with the most power in our social system are constituted of the values on the right side of Bernstein's dichotomies. As dominant discourses, they turn against marginalized discourses and create social conditions that further impoverish members of the dominated—i.e., the workers on assembly lines. It's important to remember that when we refer to a working-class discourse, we are referring to a discourse that has been created at least in part by oppressive social conditions—that is, working-class codes are not self-generated; they are generated by a social structure in which working-class people are made to think that rules have material reality: the law is the law.

Because they are privileged in industrialized cultures, the values on the right side of the dichotomies are also privileged in schools—particularly as one goes up the grade levels (Brice Heath 1983; Bourdieu 1984; Rose 1989). The fact that that the students who succeed in higher education are disproportionately from the dominant classes implies that they exhibit these values and, further, that these are the values they have learned in their homes and social groups.[1] This correlation between academic achievement and home/social environment is accelerated because of the entitlement effect (Bourdieu 23).

A bevy of measurements reported by the National Center for Educational Statistics (NCES) imply the consequences of being born into the right or wrong social group. For example, the higher one's social class, the better one can write (that is, according to the codes that the dominant social group counts as "good" writing). Although NCES doesn't report the correlation between writing achievement and socioeconomic status (SES) per se, it uses the highest level of parents' education, which is an important indicator of one's SES. I have reproduced in table 2 a summary of average writing scores based on parents' highest education level. I have included the 2007 scores for grades 8 and 12. The highest possible score was 300.

Parent Education Level	Grade 8	Grade 12
Not a high school graduate	139	134
Graduated high school	147	141
Post high school	158	152
Graduated college	166	163

Table 2. Average writing performance of 8th- and 12th-graders, by selected characteristics of students: 2007.

U.S. Department of Education, National Center for Education Statistics, National Assessment of Educational Progress (NAEP), 1998, 2002, and 2007 Writing Assessments, NAEP Data Explorer.

The same correlation between SES and achievement held for reading, math, geography, history, and science NAEP assessments (see *National*

1. The National Center for Educational Statistics reports, for example, in 2001 that 43% of first-generation students enrolled in 4-year colleges completed their bachelors, compared to 74% of students with at least one parent with a degree beyond a bachelor's. Family income statistics show the same relationship: 50% of students with family income of $25,000 or less in 1994 graduated; 74% of students with family income of $70,000 or more (National Center for Educational Statistics, 2003).

Center for Educational Statistics, 1999, chapter two).[2] I have also checked this correlation of writing ability in several other NCES reports—the unsurprising correlation between education level and test scored is repeated in every NCES report.

The web of relationships between socioeconomic status and academic achievement involves more than outcome scores, just as success in life-after-school involves more than money. Students from higher socioeconomic groups enjoy higher self-esteem for obvious reasons (Apple 1982; Trusty et al. 1994), not the least of which, as Erving Goffman has put it, is that they have more stage props available (as cited in Kerbo 1996, 374). Children are fully aware of the adult social structure in which wealth, expensive clothes, big houses, and the other trappings of high SES are admired; consequently, the hierarchical valuing is reproduced in the children's worlds. The children who benefit from this system of admiration do well in school, participate in extra-curricular activities, and are elected to student government. The National Educational Longitudal Study (1994) found the following correlation between SES and percentage of participation in student government: In the lowest quartile of SES (calculated according to a composite score based on parental education, occupation, and income), 11% participated in student government; in the 2nd and 3rd quartiles combined, 14.7% participated; in the 4th quartile, 19.8% participated—almost twice the participation rate of the lowest quartile (National Center for Education Statistics, 1999, Table 147).

The connections between self-esteem, participation in student government, and high grades are in turn reflected in the membership of the Academic Honor Society. The lowest quartile had a 9.6% membership; the two middle quartiles had a 15.9% membership; and the highest quartile had a 29.5% membership—over three times the rate of the lowest quartile. Membership in the Academic Honor Society is of course the membership that counts. By contrast, the memberships that don't count are the Future Teachers of America (future education majors), Future Homemakers of America, and Future Farmers of America. The percentages of participation, predictably, are reversed: 24.8% in the lowest quartile, 19.7% in the two middle quartiles, and 9.9% in the upper quartile (National Center for Education Statistics, 1999, Table 147).

2. For this and the following statistics, I have used the 1999 database from the National Center for Educational Statistics because they gave a more complete breakdown of kinds of student activities than was reported in subsequent years.

The success in school leads to higher SAT scores with a steady linear progression—the higher the student's SES, the higher the SAT score. In 1997, the average score on the verbal for students with family incomes under $10,000 was 428; with incomes over $100,000, the average score was 559 (National Center for Educational Statistics, 1999, Table 136). To paraphrase, Rexford Brown's (1978) famous statement, you don't need all these tests to assess student potential; you just have to count the bathrooms in their homes.

WORKING-CLASS STUDENTS' RIGHTS TO THEIR OWN LANGUAGE

Bernstein's theory and the deficit/competence dichotomy offer a framework within which to read the 1974 CCCC statement on "Students' Rights to Their Own Language." The statement begins with this declaration: "We affirm the students' right to their own patterns and varieties of language—the dialects of their nurture or whatever dialects in which they find their own identity and style."[3] The statement can be read along an idiosyncratic or a social group axis. "Whatever dialects in which they find their own identity and style" implies the idiosyncratic axis and "the dialects of their nurture" implies the social group axis. For my purpose, the idiosyncratic axis is largely irrelevant—the writers may have been referring to usage patterns that reflected such things as age or gang groupings. Although "social group" is the writers' preferred term, social class conflicts are implied. In the third sentence, they write, "The claim that any one dialect is unacceptable amounts to an attempt of one social group to exert its dominance over another." One can imagine the discussion among the writers, say between Richard Larson, Geneva Smitherman, Richard Lloyd-Jones, and Ross Winterowd, as they hammered out a sentence to which they could agree but that would not

3. Here is the full statement: We affirm the students' right to their own patterns and varieties of language—the dialects of their nurture or whatever dialects in which they find their own identity and style. Language scholars long ago denied that the myth of a standard American dialect has any validity. The claim that any one dialect is unacceptable amounts to an attempt of one social group to exert its dominance over another. Such a claim leads to false advice for speakers and writers, and immoral advice for humans. A nation proud of its diverse heritage and its cultural and racial variety will preserve its heritage of dialects. We affirm strongly that teachers must have the experiences and training that will enable them to respect diversity and uphold the right of students to their own language. (http://www.ncte.org/ccc/ex.html).
Richard Larson was the chair of the CCCC committee on language; the other members Melvin Butler, Adam Casmier, Ninfa Flores, Jenefer Giannasi, Myrna Harrison, Robert Hogan, Richard Lloyd-Jones, Richard Long, Elizabeth Martin, Elisabeth McPherson, Nancy Prihcard, Geneva Smitherman, and Ross Winterowd.

alienate the conservative members of the discipline.[4] The sentence is open to a range of interpretations, including the imposition of a variety of adult dialects on that of grade school children or teenagers. But within Smitherman's mind was the imposition of the white middle-class dialect on Black English (see Parks 2000, 110-111). Others, as well as Smitherman, may have also been thinking—as I am—of the imposition of middle-class English on working-class speakers. Such an imposition, the writers claim, "leads to false advice for speakers and writers, and immoral advice for humans."

From the statement itself, one has to imagine the nature of this false and immoral advice—it might go like this: "I speak and write better than you because I speak and write correctly, and you don't. My correct language also makes me a smarter and better person than you." The writers were clearly worried about English teachers who were offering this kind of advice in one guise or another. The rationale for offering this advice returns us to the deficit/competence dichotomy—the "correct" way is to write in elaborated code which allows one to make oneself explicit through text. This is also the code that is generally a consequence of having been raised in an individual/open/position-oriented family, i.e., the middle and upper classes. Here is how the writers of "Students' Rights" phrase the dilemma:

> We have also taught, many of us, as though the "English of educated speakers," the language used by those in power in the community, had an inherent advantage over other dialects as a means of expressing thought or emotion, conveying information, or analyzing concepts. We need to discover whether our attitudes toward "educated English" are based on some inherent superiority of the dialect itself or on the social prestige of those who use it. We need to ask ourselves whether our rejection of students who do not adopt the dialect most familiar to us is based on any real merit in our dialect or whether we are actually rejecting the students themselves, rejecting them because of their racial, social, and cultural origins. ("Background Statement")

As Bernstein (1971) points out, there is more to the issue than mere rejection—there is also a system of privileging children who come from homes in which the "power" dialect is spoken, in which the "power"

4. The writers were trying to be particularly sensitive to the lower class status of writing teachers within English departments, *manned* primarily by teachers for whom middle-class correctness was God's language. The writers were not concerned with the public reception (Richard Lloyd-Jones, personal communication).

cognitive habits are practiced, and in which the "power" social environment prevails. But when applied to social class, the fundamental question is clear: do the values on the left side of Bernstein's dichotomies describe a deficit English? Or do we interpret it as deficient because of who speaks it?

Arguments for the superiority of middle-class English at first blush seem sensible—we have heard them many times even from leftists like James Berlin, whom we have known as a defender of marginalized social groups. The superiority of middle-class English is embedded in Berlin's (1987) fundamental premise of social epistemic rhetoric: "Knowledge does not exist apart from language" (166). Or as he paraphrases Michael Leff: "knowledge is itself a rhetorical construct. . . . All reality, all knowledge, is a linguistic construct" (165). One can infer from these claims that sophisticated knowledge and sophisticated language are dialectically related. More specifically: the more names one has for different shades of the same general category, the finer one's ability to make distinctions—and the ability to make distinctions leads to more names. Likewise, with relationships and syntax.

There are many implications from the theorized inextricability of thought from language. They have in fact been so frequently cited in the past several decades that the rhetoric and composition community seems to take them for granted. As a consequence of the work by Havelock (1963) and Ong ([1982] 2000), we have learned, for example, to privilege writing over speech. The basic rationale is that writing allows authors to exteriorize their thoughts where they can be analyzed, refined, classified, organized—all sorts of wonderfully academic exercises that lead to higher level thought. The next move lies in Berlin's (1987) paraphrase of Harold Martin: "Since thought is language, . . . students will learn to write in order to improve their thinking" (168).

The logic marches inexorably toward this conclusion: writing teachers must teach their students how to write in middle-class English because that is the dialect that can accommodate sophisticated, abstract thinking. It is the dialect in which meaning is made explicit rather than left in the gaps about which teachers are forever complaining. Some people might even argue that people in the dominated classes are dominated because of their language: without middle-class English, they are limited in their abilities to generalize from instances and manipulate the relationships among the generalizations. They are stuck in the concrete and linguistically incapable of going beyond gross distinctions.

As a consequence of this theorized superiority of middle-class to working-class English, the former has been called the language of power. There are two sources of power: one, it's the language that allows power discourse; two, it's the language of powerful people. Richard Ohmann claims this latter warrant for teaching middle-class English: students need "to master standard written English . . . to become capable of participating in a linguistic community of considerable importance in our culture" (1964, 21). Although Ohmann also insisted that students should study other dialects as dialects rather than incorrect English, middle-class English still gets sorted out as the language of power because only those who speak and write it get a chance to become makers and shakers.

David Bartholomae and Anthony Pretrosky (1993) offer an extension of Ohmann's argument by constructing a rationale for initiating first-year students into the academic community. Academic discourse in this argument is a subcategory of middle-class English—more accurately, it is a subcategory of the language that the intellectual fraction of the middle classes speak and write. Students are supposed to learn how to write in this more powerful discourse by reading it, working with and against it, and practicing it in verbal sword fights with the masters. Bartholomae and Petrosky's rationale spawned a new reason for teaching students how to write in academic discourse—through practicing these kind of academic moves, students will learn how to think critically, implying that most students who haven't had the benefit of a required writing course don't know how to think critically because they lack the moves and the language capable of containing them. This pedagogy positions professors or public intellectuals as the initiated who through their essays or instruction will help the uninitiated into new ways of seeing. If we follow the logic of the deficit thesis that stretches from Bernstein through Berlin to Bartholomae and Petrosky, we can infer that students who are most in need of initiation come from the working-classes because their language, environment, and cognitive habits are antithetical to current notions of critical thinking.

So we have a series of arguments that answer the "Students' Rights" committee's question: ". . . whether our rejection of students who do not adopt the dialect most familiar to us is based on any real merit in our dialect or whether we are actually rejecting the students themselves, rejecting them because of their racial, social, and cultural origins." If we follow the logic of the deficit thesis implied in Bernstein's, Berlin's, and Bartholomae and Petrosky's argument, we can infer that working-class

English is a deficit language. We are rejecting it not because of who speaks it but because it does not allow for sophisticated thinking. The point, after all, of claiming that knowledge is a linguistic construct is, in the final analysis, that those of us who have better linguistic constructs have better knowledge. No one should be surprised that middle-class writing professors (even those with working-class origins) are drawn to the social epistemic position on the relationship between language and knowledge. I am myself.

MIDDLE-CLASS ENGLISH AND FRESHMAN COMPOSITION

I want to look closely at an elaborated answer to the "Students' Rights" question of whether we are teaching "educated English" because of "some inherent superiority of the dialect" as measured against the deficit dialect of working-class English. Although Lynn Bloom (1996) validates her answer to this question on the basis of "middle-class values" rather than cognitive operations (656), her argument in "Freshman Composition as a Middle-class Enterprise" is still the same as Berlin's (1987), Bartholomae's (1985), and Bernstein's (1971): when push comes to shove, our language is better than yours, and if you want to get into the game, you had better learn ours.

Bloom's title signals that she isn't going to mince words. She begins her article with the announcement that "Yes, freshman composition is unabashedly a middle-class enterprise. . . taught by middle-class teachers in middle-class institutions to students who are middle-class either in actuality or aspiration" (655). Bloom is frank—although not entirely accurate: from the results of my survey (see footnote 5), approximately 1% of postsecondary writing teachers in English departments come from the upper-class, 69% from the middle classes, and 30% from the working classes.[5] On the other hand, Bloom is essentially right: although many of us came from the working class, we *are* middle-class—we aspired to the middle-class enterprise.

Reiterating Bartholomae and Petrosky's argument (1993), Bloom wants to enable students "to think and write in ways that will make them good citizens of the academic (and larger) community, and viable

5. I am being generous with the notion of working-class, including in it families who may be earning a significant income (the top primary salary being roughly equivalent to as much as $30,000 in 1984 dollars—i.e., a few thousand more than I was hired for as a professor in 1991). This more generous interpretation was based on parents' education (neither parent had a college degree). A stricter interpretation based solely on income would put the working-class academic figure at about 12%.

candidates for good jobs upon graduation" (655). The deficit argument is implied when Bloom explains that middle-class English helps students learn how to think, become good citizens, and get good jobs because it embodies middle-class values: "read promotion of the ability to think critically and responsibly, and the maintenance of safety, order, cleanliness, efficiency" (655). After an ironic nod to the apotheosis of virtue—Benjamin Franklin—Bloom gets to the point:

> Indeed, one of the major though not necessarily acknowledged reasons that freshman composition is in many schools the only course required of all students is that it promulgates the middle-class values that are thought to be essential to the proper functioning of students in the academy. When students learn to write, or are reminded once again of how to write (which of course they should have learned in high school), they also absorb a vast subtext of related folkways, the whys and hows of good citizenship in their college world, and by extrapolation, in the workaday [meaning middle-class workaday] world for which their educations are designed to prepare them. In this . . . middle-class enterprise, the students' vices must be eradicated and they must be indoctrinated against further transgressions before they, now pristine and proper, can proceed to the real business of the university. Like swimmers passing through the chlorine footbath en route to plunging into the pool, students must first be disinfected in Freshman English. (656)

To set the scene more completely (in case anyone hasn't quite caught the nature of the disinfection), Bloom paints a picture of her childhood as the daughter of a University of New Hampshire professor in a town in which all "the children who could walk to school were from faculty families." The faculty kids are contrasted to the bus kids—"who lived beyond the two-mile limit and couldn't participate in extracurricular activities." Bloom then makes a gesture that Freire ([1970] 1995, 26) would call false charity. She describes Weldon MacDonald "our class's *natural* [my italics] leader." No one minded, Bloom tells us, that Weldon's boots smelled of manure and his jeans of kerosene, because he was handsome, smart, artistic, and athletic [i.e., a well-done son of Old MacDonald, one could say]. He was sexy, too—kissing "older girls in the cloakroom, seventh and eighth graders whose developed figures made them *bust* [my italics] children" (657). Despite the manure on his boots and kerosene on his clothes, Weldon was clearly a boy with a future. But he didn't go on to high school because he had to work on the family farm. This, Bloom writes, "seemed incredibly sad" (657).

I am thinking of how a farmer would interpret Bloom's gesture of sympathy. She sadly imagines Weldon's agrarian future against the future of Bloom and her friends. Weldon would perhaps get a girl pregnant at about sixteen. Closet-kissing was extracurricular training for fast family action. This closet-kissing and early fatherhood is simply the kind of things working-class kids do. To prepare for this role, working-class girls in Bloom's narrative are also prematurely big-busted. One can't imagine that any of the "walkers"—the children of faculty families—were big *bus*ted. Bloom and her girl friends weren't allowed to sex out early. They had "pin curled hair, white Peter Pan collars, and full skirts that reached to the tops of our bobby socks"(657-58), as if disallowing their skin to show. Bloom's narrative naturalizes their metamorphosis into college students and subsequently into "middle-class teachers."

This leaves us with Bloom writing this middle-class article about, in essence, disinfecting other Weldons by teaching them middle-class English within which lie middle-class virtues. In the rest of the article, Bloom expands on these virtues and their relationship to middle-class English. Here are the virtues:

Self-reliance, responsibility

Respectability

Decorum, propriety

Moderation and temperance

Thrift

Efficiency

Order

Cleanliness

Punctuality

Delayed gratification

Critical thinking

Although I might quibble about a few of these virtues, I believe that most of them are desirable in both writing and moral conduct. I take exception, however, to the implication that these virtues belong *only* to the middle class—and that the working classes are defined by their lack. I suspect that Bloom would shrink from acknowledging this implication—in fact, I believe it is her submergence within her social class that keeps her from seeing it—but to me as a working-class academic with a rural background, the subtext leaps off the page.

In her elaboration on "Cleanliness," she gives the subtext voice. Cleanliness, she writes, "is next to godliness in the middle-class pantheon. Dirt, like disorder is a privilege of the filthy rich and the slovenly poor" (664). A page earlier, she had claimed that "disorganized writing is as disreputable as *disorderly* [my italics] conduct, for it both implies mental laxity and shows disrespect for one's readers" (663), knitting the lower classes with disorderly housework, conduct, thinking, and writing—with the implication of punishment by incarceration or fine. The opposite is middle-class cleanliness, which leads to clear thinking and writing.

By claiming that a characteristic such as cleanliness is the property of one social class, Bloom implies that it is not a property of the infected classes. She marks the middle-class cleanliness off more explicitly when she notes that the extreme upper and lower classes are unclean. The problem with the disorderly classes is ironically accentuated by Bloom's neatness or her middle-class concern with appearances (see Bourdieu 1984, 252-53). Because parallel constructions are a mark of middle-class prose, she felt compelled to balance "filthy rich" with "slovenly poor." In addition, she could not help but pun with "filthy," which led to a double entendre with "slovenly"—is "slovenly" a restrictive or non-restrictive adjective? There is of course no way to interpret her intent, but the possibility of the non-restrictive use of "slovenly" is fed by the middle-class trope locating the poor as unclean, their houses unkempt, their yards unmown with junk and old cars scattered about. Weldon is only a slightly modified version of this trope.

The deficit argument lies within each of Bloom's elaborations on the pantheon of middle-class virtues. For instance, one can't help but think of Weldon kissing seventh graders in the closet when Bloom refers to the *delayed-gratification* virtue of the middle class. More explicitly, she says, "It is a middle-class virtue to work and scrimp and save in the present for long-term gains in the future—such as the fruits of an education . . ." (665). Although Bloom may not have been referring to Weldon, there Weldon is, a silent embodiment of the working-class—shit-booted and kissing D-sized girls in the closet—a dark, private place where other things might happen. In Bloom's middle-class vision, the Weldons of the world need to be disinfected through the agency of middle-class prose.

Bloom implies that students who do not learn to "correct" their "disorganized writing" are guilty, like the poor, of "mental laxity," underwriting a social theory that explains why the poor are poor: they are dependent, irresponsible, disreputable, immoral, indecorous, immoderate, extravagant, inefficient, disordered, slovenly, unpunctual, incapable of delaying gratification, and naïve thinkers. It is perhaps moot that I could argue that many members of the middle classes are not all that self-reliant (I don't, for instance, know many middle-class professors who repair their electricity or plumbing), decorous, thrifty, efficient, and so on. What we have here is a middle-class version of how the middle classes would like to see themselves—and how they would like to see the working classes who are everything the middle classes are not, a vision that explains why the poor are poor.

The argument works to sanitize a middle-class bias. Bloom puts the finishing touches on the argument by shifting her terms: middle-class English becomes "normative discourse" (659), which later elides into the "lingua franca for writing in the academy," a "normative view of language," and finally back to "standard English" (664). So Bloom's argument is one more direct answer to the question posed by the Committee on Students' Rights to Their Own Language—we should teach "standard English" because it is a superior dialect. Although there are moves to respect the other dialects like working-class English, the real difference between middle-class and working-class English lies not in their lexicon, syntax, or punctuation, but in the presence or absence of the virtues within them.

Bloom's is a beguiling version of the deficit argument. Her argument is structurally identical to Berlin's (1987) and Bartholomae's (1985). Group A has the correct language and kind of thinking, Group B doesn't. Group A's job is to get Group B to write and think like Group A. Group A thinks its language and thinking constitute the language and thinking of power. Group B's is the language and thinking of the disempowered. Successful socialization occurs when Group A manages to convince Group B that its language and thinking are weak.

This pattern is common among different groups struggling for social power. The identity of the groups may vary: Group B may be the laity, students, the working-class, native Americans, or peasants. In the case of Berlin, Bartholomae, and Bloom, Group B is nominally constituted of students—and by extension, other uneducated people who haven't had the benefit of instruction in Group A's required writing course, modeled on some version of the initiatory pedagogy outlined in *Ways of Reading* (1993). A plethora of research (see, for example, Alspaugh 1992; Kruse 1996) shows that working-class students are more in need of Bloom's notion of disinfection than middle- and upper-class students. Research by Bernstein (1971), Brice Heath (1983), Bourdieu and Passeron (1990), Lareau (2003), Gee (1996), and Willis (1977), make the reason for this imbalance painfully obvious—for middle- and upper-class students, instruction in the language and cognition of power represents a continuation of what they have been learning at home; for working-class students, it represents not only a break but also a contradiction—they have to choose between becoming an ear 'ole or remaining one of the lads. As an ear 'ole, they will learn to hear their parents' language with ears that condemn it. As lads, they retain their parents' language and confirm their parents' values. The lads' responses are interpreted by middle-class teachers as resistance, evidence of uneducability, or as a learning disability. The effects of privileging one group's language over another are entirely predictable. The students may have a right to their own language, but if they want to get on in our middle-class world, they had better learn ours.

4
CRITICAL THINKING

Writing teachers have been trained to believe in a necessary dialectic between language and thought—more particularly, in the dialectic between written language and thought. The line of thinking runs roughly from Vygotsky ([1962] 1975), through Havelock (1963) and Ong ([1982] 2000) to Berlin (1987). The gist of this literacy trope is that as you speak, so you think—and even more so, as you write. The pull of this trope is so strong that it has usually gone unquestioned among writing teachers—with the notable exception of Stuckey's (1991) virulent attack in The Violence of Literacy.

The dialectic between written language and thought was the foundation of Bloom's (1996) argument for teaching middle-class virtues by way of middle-class English. Bloom reserved as the last virtue "critical thinking," the trump card of middle-class English. Although we may be a little unclear about what we mean by "critical thinking," it's what we're all for. To say you are teaching your students how to "think critically" is like waving the flag on the 4th of July. We include it in our mission statements, in our syllabi, and in the rhetorics and readers we have our students purchase—many of which struggle to get critical something or other in their titles. It's one of the five major objectives of the WPA outcomes statement and a guarantee of grants. The attractive feature about critical thinking is that teaching it implies you know how to do it. It's almost unthinkable to challenge it as one of our major purposes of instruction.

But that's what I want to do here. I at least want to challenge the critical thinking agenda as an extension of Bloom's argument for teaching middle-class English. Critical thinking is a more subtle social class signifier than language, tricking many who think they are working on behalf of marginalized social groups into acting as agents of social reproduction—or as I have suggested in my title, going north when they are thinking west. Although teachers may hope they are teaching students how to think critically about social myths that perpetuate oppression, they might be oblivious of the possibility that critical thinking, as it is frequently conceived and taught, is no more class neutral than middle-class English.

In my analysis of the relationship between critical thinking, writing instruction, and social class reproduction, I will divide the critical thinking agenda into cognitive and social strands. The cognitive is derived from the fields of psychology, education, and the informal logic branch of philosophy and the social from the fields of literary criticism and the socio-economic branch of philosophy. These different origins lead to different pedagogies—and consequently, to different methods of social class reproduction. I will emphasize the ways the two strands marginalize working-class students in our required writing classes. I am not arguing that all teachers who promote either strand of critical thinking in their classrooms slip into reproductive pedagogies or that we should not teach our students how to engage in middle-class critical thinking. As students need to know the middle-class language games we play, so they need to know how to play our middle-class game of "critical thinking," but we should steer clear, as Lilia Bartolomé (1998) has argued, of a replacement pedagogy. Rather than replace, we need to give them alternative ways of thinking and writing that fit the new social situations in which they will increasingly find themselves. An additive rather than a replacement pedagogy is particularly important when we are working with students who come from marginalized social groups whose home languages and ways of thinking seem in conflict with the ones we need to promote if we expect to help our students get on.

HISTORICIZING CRITICAL THINKING

Academics have historically used "critical thinking" interchangeably with "critical inquiry." The difference in the phrases lies mostly with who uses them. In English studies, people on the literature side tend to use "inquiry," whereas people on the composition side more frequently use "thinking." These differences have evolved from the histories of the terms.

"Critical inquiry" is by far the earlier term. It first appeared in a book title in 1748, *A Critical Inquiry into the Opinions and Practice of the Ancient Philosophers Concerning the Nature of the Soul and a Future State.* Subsequent books with "critical inquiry" in their titles appeared every fifty years or so with the subjects being Alexander the Great (1793), "antient armour" (1800; 1824; 1842), teeth (1846), Scottish language (1882), medieval building (1884), and Argentine Economic History (Garcia 1973).

Its use as a trope didn't gain currency until after the inauguration in 1974 of the journal, *Critical Inquiry*. Although *Critical Inquiry* was announced as an interdisciplinary journal concerned with the "theory, method, and exploration of critical principles in the fields of literature, music, visual arts, film, philosophy, and popular culture," articles in the journal lean toward literary criticism and philosophy—particularly philosophy concerned with literary production. The journal's focus on critical principles suggests a reference to Wimsatt and Beardsley's ([1946] 1971) article, "The Intentional Fallacy," in the last line of which they claimed that looking for meaning outside the text would not be "critical inquiry," by which they mean the kind of inquiry that critics do, i.e., using the tools available to critics but unavailable to those untrained in textual interpretation.

After the inauguration of *Critical Inquiry*, use of the phrase accelerated in comparison to its use in the previous three centuries. Three books from the years 1974 to 2001 included "critical inquiry" in their titles. One was about economic history (1987), one about psychoanalysis (1988), and one about politics (1993). The phrase appeared in only 11 titles in the MLA electronic database, which tracks literature-related articles, books, and dissertations from 1963. The phrase was more popular in education-related articles, books, and dissertations, appearing in 20 titles in works reported in ERIC after 1965.

Although a Johnny- or Jill-come-lately, "critical thinking" is by far the more popular phrase. It first appeared as part of a book title in 1941, *Experiment in the Development of Critical Thinking*. Its author, Edward Glaser, later created with Goodwin Watson the most enduring critical thinking test in the business—the *Watson-Glaser Critical Thinking Appraisal*. In 1946, Max Black published *Critical Thinking: an Introduction to Logic and Scientific Method*, territorializing philosophy and logic as the appropriate field for investigating critical thinking. After the second edition to Black's book in 1952, books with "critical thinking" in their titles began to appear every few years. From 1980 to 1983, the pace accelerated with one or two appearing every year or so. In 1984, the title appeared in 8 books, a rate that remained steady until 1989, after which the rate steadily accelerated to about 25 a year from 1995 until the present with the rate remaining strong. This increased popularity was likely stimulated by the series of conferences on critical thinking at Sonoma State University, the first of which occurred in 1980.

A scan of the references reported in ERIC since 1966 dramatizes the degree to which "critical thinking" overshadowed "critical inquiry." To the 20 education-related publications with "critical inquiry" in their titles, there were 1072 with "critical thinking." The burst in education-related publications occurred a few years earlier than with books, a consequence of publication lag time. Through the 60's and 70's, the publications including "critical thinking" in their titles occurred at the rate of 8-10 a year. But in 1983, this rate accelerated to 23 in a year. It was 24 in 1984, 41 in 1985, 58 in 1986, 72 in 1987, 65 in 1988, and reached an all-time high of 83 in 1989. Thereafter, the frequency was about 60 a year with some diminishing toward the end of the 90's when it steadied to a rate of about 50 a year. Publications reported in MLA were more sparse—only 24 were reported in the years from 1963 on; 14 of those were dissertations. Literature specialists clearly have not bought into critical thinking. As we will see, critical thinking is more popular with progressive writing teachers when it's coupled with social reform, and it's a hit in education.

The meaning of the two phrases is unproblematic at a high level of generalization. But at a lower level specifying the process and the subject of the inquiry, differences emerge. The differences become more pronounced when one considers how or whether one can teach students to think or inquire critically.

Most writers who use either term agree that critical inquiry or thinking involves an informed look at something. Writers usually refer to examining available information and drawing intelligent conclusions through analyzing and synthesizing, preferably from a relatively disinterested point of view. The first catalogued book (1748) with "critical inquiry" in its title surveys the available information and takes a stance on whether the ancient philosophers believed in an afterlife and soul. The tone of the text is aggressive as the writer presses his point, but by the beginning of the 19th century, the writer assumes a more disinterested tone. The author of *A Critical Inquiry into Armour*, Samuel Rush Meyrick (1800), carefully examines old pictures, poems, and miscellaneous writings to infer what armor in a particular time and place looked like. I suspect the subsequent critical inquiries into teeth, language, economics, and so on carried forward this notion of being able to examine the available evidence and from it formulate careful conclusions.

This inquiry is the kind of scientific investigation John Dewey ([1938] 1963) theorized as the foundation of progressive education. Students were to be taught how to investigate phenomena without prejudice, relying on what they could discover rather than on what they had been told. The reference to "critical inquiry" in Wimsatt and Beardsley's ([1946] 1971) article, "The Intentional Fallacy," adds to the scientific investigation a reliance on the tools of the trade. Critical inquirers had to learn how to employ the tools of critics, how to look carefully at an object (a poem) and establish through the use of critical strategies the meaning that lay in the object, the poem. Speculation and fancy about what the writer had meant were not a part of the critic's conversation.

COGNITIVE STRAND

The cognitive strand of critical thinking theory developed from Edward Glaser's 1941 book, and Robert Ennis's (1962) landmark essay "A Concept of Critical Thinking." Ennis's thumbnail definition of critical thinking is frequently referred to as the key to critical thinking. Critical thinking, Ennis wrote, "is the correct assessing of statements" (83).

Ennis's focus on evaluating the "correctness" of statements is a consequence of his indebtedness to a previous article, "The Improvement of Critical Thinking," by Othanel Smith (1953). Ennis quotes Smith's statement that "if we set about to find out what . . . [a] statement means and to determine whether to accept or reject it, we would be engaged in thinking, which, for lack of a better term, we shall call critical thinking" (83). Although Ennis acknowledges that critical thinking may apply to value statements as well, the primary subject in his essay is truth statements. Whereas Smith distinguishes between good and bad thinking, Ennis transforms the "good" into "critical" and the bad into non-critical thinking. Ennis also contributed to the nascent discussion of critical thinking by abstracting out of the multiple possibilities the following twelve-step program for assessing the correctness of statements:

1. Grasping the meaning of a statement.
2. Judging whether there is ambiguity in the line of reasoning.
3. Judging whether certain statements contradict each other.
4. Judging whether a conclusion follows necessarily.
5. Judging whether a statement is specific enough.
6. Judging whether a statement is actually the application of a certain principle.

7. Judging whether an observation statement is reliable.
8. Judging whether an inductive conclusion is warranted.
9. Judging whether the problem has been identified.
10. Judging whether a statement is an assumption.
11. Judging whether a definition is adequate.
12. Judging whether a statement made by an alleged authority is acceptable. (84)

Ennis carefully details several sub-steps for each of the steps above. He also qualifies his analysis by noting that a critical thinking program is constituted of concepts abstracted from the far messier business of real-time responses to statements. In subsequent articles on critical thinking, this qualification is usually ignored.

Black's ([1946] 1952) *Critical Thinking: An Introduction to Logic and Scientific Thinking* deserves mention in an overview of the cognitive strand. Although Black's book is rarely referred to in subsequent literature, it is a detailed precursor of Ennis's (1962) article and a logician's version of Smith's (1953) statement. Black devotes the first section of his book to a logical analysis of statements, the second to language, and the third to induction. These sections correspond to three of the important dimensions one has to consider when thinking critically about any kind of discourse. One has to consider the relationships between elements of the discourse, the language in which those relationships are expressed, and the empirical evidence supporting any claim. To these three, one might add—as later writers did—the social, historical, and field contexts.

Black, Smith, and Ennis have several criteria in common that shape later explanations of critical thinking. First, they are all concerned with the "correct" interpretation of statements—correctness here is understood as a description of a statement's truth-value, i.e., the extent to which the audience is justified in believing it. The focus on language stems from the post World War II concern with propaganda. The point of critical thinking was to guarantee that citizens would not be swayed by propaganda. The emphasis on truth statements might be a consequence of these writers' professions and social class—that is, in the working classes, speakers might be more interested in instrumental statements: "Put burlap over the cement and hose it down at the end of the workday." At the end of his article, Ennis includes an interesting statement linking his explanation of critical thinking to social class issues. When predicting various correlations, he says, the "social-class status of students outside of the upper-class, is probably correlated with *critical thinking*" (108). He doesn't elaborate on this statement, but the tenor of the article implies that with the curious exception of the elite, the higher the social class, the more critical the thinking.

This implication is interesting for two reasons. First is the exclusion of the upper-class, echoing Bloom's (1996) interpretation of the "filthy rich" (664). Perhaps Ennis assumes that children of the upper-class don't have to work hard to maintain their social positions; consequently, they would likely prove the exception to his prediction. Second is his speculation that working-class students don't think as critically as middle-class students. He may be right—but only as he is defining critical thinking. Ennis implies his definition is neutral, but it is biased toward privileging the kind of thinking middle-class people do and the kind of subjects about which they argue. It is nearly a tautology to suggest that people from that social class will probably be better at demonstrating the skill that people from that social class practice.

Ennis's description of critical thinking has defined the concept of critical thinking for cognitive strand theorists. Truth statements have become the object; and deductive logic, semantic analysis, and inductive logic have become the modes of thinking critically. Most descriptions of cognitivist-oriented critical thinking is directed at argumentative discourse to the exclusion of the many other kinds of kinds of discourse about which one could think critically. Barry Beyer collected in 1985 what he interpreted as the most authoritative descriptions of critical thinking. In addition to Ennis's, we have the following elements in the eight lists of critical thinking skills cited:

> Evaluating evidence, drawing warranted conclusions (Dressel and Mayhew 1954);
>
> Evaluating reliability of authors, checking accuracy of data (Fraser and West 1961);
>
> Determining . . . accuracy of inference, deducing conclusions, evaluating strength of an argument (Watson and Glaser 1980);
>
> Determining if a statement follows from its premises, a hypothesis is warranted, a theory is warranted, an argument depends on equivocation, a reason is relevant (Ennis 1982);

> Distinguishing between statements of fact and statements of opinion, determining the difficulty of the proof (Morse and McCune, revised by Brown and Cook 1971);
>
> Separating statements of fact from statements of value, distinguishing hypothesis from evidence, recognizing logical inconsistency in arguments, distinguishing hypothesis from warranted conclusions, recognizing logical fallacies, recognizing persuading techniques (Fair 1977);
>
> Evaluating a line of reasoning, weighing evidence, identifying ambiguous statements, identifying equivocal statements. (Hudgins 1977)

With the exception of Fraser and West's list, which is more generally directed at informative writing, argumentative writing is the assumed supra-genre. My point in this discussion has been that in spite of Ennis's caveat, the cognitive strand of critical thinking has focused on only one of the categories of the aims of writing—persuasive—as James Kinneavy (1969) has defined them. Thinking critically about writing classified according to the other three aims—expressive, referential, and literary—is entirely ignored; that is, one couldn't think critically about one's diary, *The Voyage of the Beagle*, or *All the King's Men.* And thinking critically about non-discursive reality (like music, dance, farming, or engineering) is barely a footnote in the cognitive strand.

Field Dependency

The major players in the critical thinking industry are its originators: Watson and Glaser and Ennis, who with Jason Millman developed the *Cornell Critical Thinking Tests* (first iteration in 1966). Ennis, Watson, and Glaser are certainly scholars with integrity, but their vested interest in their research must be acknowledged. Their subject positions would likely incline them to welcome and promote research that substantiates the foundations for their assessments. Many researchers seem to have uncritically adopted these tests as pre and post measures of control and experimental groups. The citations for this kind of research are overwhelming. The central assumption behind the cognitive strand is that critical thinking is a generic activity that can be abstracted from instances of thinking about something. The forms that are abstracted, i.e., the moves that constitute critical thinking, can be studied and taught as an existential cognitive activity. John McPeck (1981) argues, however, that thinking is always thinking about something, even if one is thinking about thinking. He claims that critical thinking is field-dependent and consequently does not exist as a subject that can be abstracted, taught, and tested, undermining the foundation on which tests like the Watson-Glaser and Cornell assessments are built.

McPeck tracks the evolution from defining critical thinking as formal logic (Black [1946]1952) to defining it as informal logic. The connection between the two logics lies in their strategies of abstracting principles from contingent situations. Formal logic, notably in its rules for truth tables, attempts to evaluate arguments by reducing statements to symbols, like P or ~P. Informal logic attempts to do the same thing without symbols. Informal logic, for example, attempts to catalogue all fallacies. This attempt is an act of abstraction, of finding single descriptions that apply to multiple discourse acts. The search for a concept of critical thinking is a similar attempt to find principles of critical thinking that one can apply to all instances in which thinking is required.

McPeck compares the desire to discover the principles of critical thinking to the phonics movement, and argues that because critical thinking doesn't occur in a vacuum, it cannot be taught in a vacuum any more than reading can. Reading is reading about something—and one must have an initial comprehension of the subject in order to make sense out of the words one is sounding out. Teaching phonics is teaching only the sounding out. One could very well sound out words, McPeck says, but without understanding the words, one could not call it reading. Similarly, critical thinking taught apart from critical thinking about something can lead to bad teaching—and bad testing. People end up testing something that doesn't exist—or perhaps they think they are testing one thing, but they are really testing another.

"The best assessments of arguments," McPeck (1981) claims, "usually come from people with the most information about a subject and not from those merely skilled in argument analysis. In a world of complex facts, events and ideas there simply is no short cut to analyzing arguments apart from understanding these complexities" (93). McPeck notes Stephen Toulmin's claim that the uses and forms of argument are as varied as life. "Arguments within any field," Toulmin writes, "can be judged by standards appropriate within that field, and some will fall short; but it must be expected that the standards will be field-dependent, and that the merits to be demanded of an argument in one field will be found to be absent (in the nature of things) from entirely meritorious arguments in another" ([1958] 2003, 235).

I heard a striking example of the relationship between field and argument when I was researching genre acquisition in corporate environments. An executive vice president of a large Midwestern insurance company told me that when corporate executives make decisions (we were talking about an impending decision, involving millions of dollars, to switch to a new software program to track investments), they tend to value testimonials more than research. He explained that executives know how data can be manipulated, so they can't spend their time ferreting out the fudged from the unfudged data; but if they hear from X, Y, and Z that such and such policy or program has worked for their companies, those testimonials mean more than data. The vice president explained that this is because the executives know each other, so they value the word over the data. This reliance on testimonials is the flip side of what is valued in academic communities. In academic research, we discount testimonials. Presented the same evidence, then, members of the different communities would value the evidence differently. Toulmin would explain this as a difference in what counts as acceptable warrants in the different communities.

If McPeck's and Toulmin's theories of field-dependency are sound (see also Frank Smith for a critique of a field-independent theory of critical thinking), people who become expert in their version of a field-independent critical thinking may be tempted to apply the standards of critical thinking in one field to the standards of another, which could be like playing baseball in an antique shop. Richard Paul (1993), one of the leading theorists of the cognitive strand, has posted an essay, "Why Students and often Teachers Don't Reason Well," that is a ironic example of faulty assumptions based on transferring the standards of one field to the standards of another.[1]

Paul, a philosopher, takes it upon himself to show how some writing experts don't know what they are talking about in their assessment of 8th grade writing skills. Paul's primary target is Charles Cooper, co-editor with Lee Odell (1977) of one of the best known books on writing assessment. Paul's evidence is an essay Cooper and other California secondary

1. It used to be posted, that is, on http://www.criticalthinking.org. You can now purchase it for $5.00 from the critical thinking website.

writing teachers (including Marilyn Whirry, National Teacher of the Year 2000) chose as exemplary but that Paul argues should have been an example of failure because the student writer didn't exhibit critical thinking skills—as Paul defined them.

Paul makes several mistakes that would be immediately apparent to writing teachers but perhaps not to a philosopher, particularly one who has devoted most of his career to promoting a generic version of critical thinking. For example, he quotes the following brief vignette that the writer uses to introduce her evaluation essay (the writer is explaining why she likes pop rock music):

> 'Well, you're getting to the age when you have to learn to be responsible!' my mother yelled out. 'Yes, but I can't be available all the time to do my appointed chores! I'm only thirteen! I want to be with my friends, to have fun! I don't think that it is fair for me to baby-sit while you run your little errands!' I snapped back. I sprinted upstairs to my room before my mother could start another sentence.

Writing teachers will recognize in this paragraph a narrative gambit to get the reader's attention, one in fact that the writer has most likely been taught. For 8th grade writing produced in fifty minutes, most of us would probably give this writer some credit, but Paul (1993) complains, "It is clear that in this segment there is no analysis, no setting out of alternative criteria, no clarification of the question at issue, no hint at reasoning or reasoned evaluation."

Paul's failure here is that he imagines the teachers were assessing critical thinking as a field-independent, generic activity when they were in fact assessing students' abilities to write in different genres—in this case the genre of evaluation. Paul is mislead into thinking that he could apply the rules of philosophic discourse to writing instruction. For his own purposes (masquerading as an objective evaluation), Paul was judging an apple rotten because it wasn't an orange. Paul's negative critique of writing teachers like Cooper and Whirry is evidence substantiating McPeck's central claim that critical thinking, like argument, is field-dependent. Someone who thinks critically in one field may try to use the same strategies in another field and demonstrate the opposite of what he or she intends. The real mistake lies in thinking that because one may be a "critical thinker" in one field—like informal logic—that he or she is a critical thinker in all fields—like evaluating 8th grade writing in specific genres.

Intersections

Having argued that critical thinking is field-dependent, I want to hedge. There are a few features that seem common to most concepts of critical thinking. McPeck states the core of critical thinking by saying

> Let S = individual
>
> Let CT = critical thinker
>
> Let X = a field
>
> Let E = what S knows about X
>
> Then S is a CT in X if S can do X by imagining ~E or ~any subset of E. (9)

The formula implies that people are thinking critically when they are suspicious of all they know on a given subject for the purpose of theorizing alternative ways of interpreting a situation. McPeck also describes critical thinking as an "appropriate use of reflective skepticism" (7). "Appropriate" and "reflective" are key words. Reflective implies a moment of stepping away from the situation to think again. This is different from knee-jerk skepticism. People who are habitually skeptical have not realized that reflective skepticism has to be appropriate: when the pitch comes and the ball is headed toward one's head, it's a bad time to question the material reality of the ball. McPeck points out that knowing *when* to explore alternative realities is also a function of one's knowledge of the field, X. It then follows that the "appropriate" criterion of critical thinking is also field-dependent. Thinking critically while playing baseball is different from thinking critically as a banker. Experts in each will know when to explore alternative realities and when to act—in their own fields.

SOCIAL STRAND

The cognitive strand, with its roots in education, psychology and logic, focused primarily on argumentative discourse. The social strand, with its roots in literary criticism and socioeconomic philosophy, focuses on social structures. Influenced by literary theory, theorists in the social strand use "critical inquiry" interchangeably with "critical thinking," referring to a disinterested, informed investigation of a subject with some implication of using the right tools for the right job. The term crosses over in several instances, sometimes within the same article or book. In "Using a Critical

Inquiry Perspective to Study Critical Thinking in Home Economics" Jane Plihal (1989) traces critical inquiry back to Hegel's and Marx's influence on the Frankfurt school in the 1920s. The tools and frame of mind that the literary theorists use to inquire into texts are the same tools and frame of mind the critical theorists from the Frankfurt school use to inquire about social reality. One learns to "look again" with de-socialized eyes or at least with eyes that take into account the degree of the viewer's socialization (see also Bourdieu 1984, 12, 482; Kuhn 1970). Plihal quotes Marx on the necessary connection between criticism and desocialization:

> [we] wish to find the new world through criticism of the old; . . . even though the construction of the future and its completion for all times is not our task, what we have to accomplish at this time is all the more clear: relentless criticism of all existing conditions, relentless in the sense that the criticism is not afraid of its findings and just as little afraid of conflict with the powers that be. (40)

The purpose of the social strand of critical thinking is to deconstruct through "relentless criticism" social structures that naturalize the exploitation and oppression of dominated social groups. Narrowing the focus, Plihal (1989) says the "object of critique in critical inquiry is ideology" (40) because ideologies are the instruments through which social structures are naturalized.

Plihal notes the collaborative quality of critical inquiry—a Socratic feature that is strikingly absent in the cognitive strand of critical thinking. Theorists in the social strand assume one cannot think critically by oneself, because if one doesn't communicate with others who come from different situations and necessarily have different visions, one remains trapped in the room of one's mind. A necessary condition for critical inquiry is therefore non-distortive communication, or Habermas's "ideal speech situation." The features of Habermas's ideal speech situation (comprehensibility, factuality, sincerity, and justifiability) describe an ideal scientific community in which participants verify that they understand each other's language, that all known facts support statements, that all speakers utter only that which they believe to be true, and that only justifiable claims supported by data and appropriate warrants are entered into the discourse. Under these conditions, participants can move through dialogue toward an approximation of truth.

After setting down these conditions for critical thinking, Plihal outlines a procedure for a critical inquiry in home economics courses. The steps are as follows:

1. Identify movements or social groups whose interests are progressive.
2. Develop an interpretive understanding of the intersubjective meanings, values, and motives held by all groups of actors in the subjects' milieu.
3. Study the historical development of the social conditions and the current social structures that constrain the participants' actions and constrain their understandings.
4. Construct strands of the determinant relations between social conditions, intersubjective interpretations of those conditions, and participants' actions.
5. Elucidate the fundamental contradictions which are developing as a result of current actions based on ideologically frozen understandings.
6. Participate in a program of education with the subjects that gives them new ways of seeing their situation.
7. Participate in a theoretically grounded program of action which will change social conditions and, in addition, will engender new less alienated understandings and needs. (44)

Plihal's final step predicts the transformative agenda that drives most current critical pedagogy. The researcher and those with whom he or she is working should engage in a "continuous cycle of critique, education, and action" (45)—the final object of which is to empower both teachers and students by helping them understand they can work to transform their material conditions of existence.

Although Plihal cites D.E. Comstock's (1982) "A Method for Critical Research" as her primary source for her protocol, she could have cited Paulo Freire's ([1970] 1995) program for developing literacy among oppressed social groups. In Freire's pedagogy, critical thinking, which he calls "conscientization," is necessary for literacy development. Freire's pedagogy is generally described as critical, liberatory, or emancipatory because its purpose is to free educatees from dominated forms of consciousness. Like Plato's philosophers, they will emerge from the cave into a "more fully human" (29) condition of existence. By fully human, Freire means a condition in which people understand they have the power to transform their material conditions of existence. As in Plihal's program, dialogism is a necessary condition for conscientization: it has to be arrived at *with* and *for* others. For the oppressed, a true literacy will necessarily lead toward social action, the purpose of which is to continuously work against all forms of social oppression based on the mythologies of individualism, democracy, and equal education.

Freire's pedagogy is a blueprint for Plihal's program of critical inquiry. Before embarking on a literacy development program, Freirean educators visit the community and investigate with progressive members of the community the specific conditions of the community. The educators write up and collaboratively evaluate with their co-investigators a report in which they try to distinguish between the real and potential knowledge of community members and identify generative themes with embedded contradictions in the people's lives. The "real" knowledge—corresponding to Marx's false consciousness—is limited by the participants' uncritical perception of their living situations, a perception that has been presented to them through dominant ideologies. The potential knowledge is the world opened up by critical thinking, which Freire defines as

> thinking which discerns an indivisible solidarity between the world and the people and admits of no dichotomy between them—thinking which perceives reality as process, as transformation, rather than as a static entity—thinking which does not separate itself from action, but constantly immerses itself in temporality without fear of the risks involved. Critical thinking contrasts with naïve thinking, which sees "historical time as a weight, a stratification of the acquisitions and experiences of the past,"[2] from which the present should emerge normalized and "well-behaved." For the naïve thinker, the important thing is accommodation to this normalized "today." For the critic, the important thing is the continuing transformation of reality, in behalf of the continuing humanity of [people]. (73)

Several themes are embedded in Freire's description of critical thinking: people are made by and make the world, reality/history is mutable, thinking and action are inseparable, critical thinking leads toward the continuing transformation of the world with the purpose of enabling all people to participate in that transformation. When we take from people their power to be an active part of that transformation or the continuous re-making of their conditions of existence, we are taking from them the essential feature of their humanity, which distinguishes them from

2. Freire ([1970] 1995) includes in the quotes a phrase that he cites as from a friend's letter.

animals, which, Freire claims, only respond to the world. Transforming the world, engaging in a "program of action which will change social conditions and, in addition, will engender new less alienated understandings and needs" (Plihal 1989, 44) is the ultimate purpose of the social strand of critical thinking.

Freire's definition of critical thinking is in sharp contrast to Ennis's (1962) "the correct assessment of statements" (83) with its focus on argumentative discourse. In part, the social strand of critical thinking is an application of the cognitive strand, but critical thinking in Freire's philosophy is about much more than statements. It is about a way of seeing the world and one's relationship to it. It is what one ends up with when one applies McPeck's (1981) formula: S is a CT in X if S can do X by imagining ~E or ~any subset of E.

To McPeck's (1981) formula and his caution about "appropriate skepticism," I would add that one needs to know that one is knowing, a central condition of Freire's conscienticization ([1970] 1995, 66-67). Being reflexively aware of one's consciousness entails knowing that one is knowing from a situated position. Paul (2004), who should have been more careful practicing it, has called this "epistemological humility." Knowing that one's knowledge is situated is tantamount to knowing that it is shaped by one's experiences and culture. A critical thinker should consequently be open to multiple ways of knowing. Being open to multiple perspectives must not, however, erase one's epistemological humility—people who would like to call themselves critical thinkers tend to compliment themselves on their catholic vision, comparing themselves favorably to those with more parochial perspectives. Critical thinkers must always be aware that knowing-that-one-is-knowing is itself situated knowledge. Seduced by their social group and professional mythology, people who think they know they are knowing may be beguiled into thinking they know more than non-critical thinkers who don't know they are knowing. I have seen this happen in English departments. What I don't know is whether I have been able to see myself seeing.

Finally, critical thinking theorists agree that thinking critically entails a perception of the relationship between the local and global. To try to understand things as if they existed existentially is naïve—a key concept in the evaluation of social class language and thinking. Critical thinkers analyze the part and then speculate on its relationship to a larger structure. This is precisely what Jeannie Oakes and Kenneth Sirotnik

(1983) were arguing when they claimed that a productive analysis of the problems in education had to be predicated on understanding the parts schools play within the larger social structure. To analyze an educational problem existentially ensures misdiagnosis. It's like thinking that students are not learning to read well because they haven't learned how to sound out the individual phonemes—or that people haven't learned how to think critically because they haven't learned how to identify assumptions embedded in argumentative discourse.

To summarize: the cognitive strand of critical thinking, focusing on argumentative discourse, is more specific than the social. Although cognitive-oriented theorists acknowledge that the extent of one's knowledge in a field will be a factor in one's ability to think critically, they tend to proceed as if critical thinking were field-independent, a consequence of their attempt to abstract the concept of critical thinking from the act of thinking critically. Although the social strand of critical thinking is more general than the cognitive, theorists in this strand tend to assume critical thinking is field-dependent, involving a dialectic between thinking and acting, the self and the world. Members of the cognitive strand want to teach critical thinking. Members of the social strand want to remake the world.

5
ARGUING

Although I hesitate to adopt all features of critical thinking as goals in my required writing classes, I want my students to know that their knowledge is situated, which is a precondition of being able to see from multiple perspectives, allowing writers to read their texts (and themselves) with other readers' eyes. If I were asked to focus on one writing skill that marks the transition from high school to college writing, it would be this ability to see one's own text/self with others' eyes.

Linked to the ability of seeing from multiple perspectives is the ability to recognize the social context within which claims are interpreted as either self-evident (see Bartholome's [1985] notion of commonplaces, 42-43, and Skorozewski's [2000] refreshing article on academic clichés) or in need of substantiation. Because of their limited exposure to multiple perspectives, young writers tend to be locked into circumscribed social groups within which a set of claims are treated as irrefutable truths, such as "freedom is our god-given right" or "competition encourages higher achievement." At the university level, students are confronted with a kaleidoscope of social groups as well as with the new demands of various academic discourse communities, many of which bring with them new sets of assumptions. By being able to see with others' eyes, students gradually come to understand the contingent nature of assumptions they once believed were stable. Students who learn how to question their own assumptions will not only write more effectively but also learn how to accommodate diversity. In addition, the search for data and warrants substantiating challenged claims trains students to think more clearly, to require reasons for their beliefs and to require the same reasoning of others.

Because of their connection with effective writing, I make these features of critical thinking a fundamental part of our required writing courses. But I am less willing to adopt other features of critical thinking associated with argumentative writing—not because they are undesirable but because when they are uncritically integrated into instruction, they may function as another weeding-out mechanism acting against working-class students. By uncritically, I mean adopting them as if they were class neutral rather than loaded with attributes that make them more accessible to middle-class than working-class students.

SOCIAL CLASS AND ARGUMENT

Required writing programs commonly sequence instruction by having students work in various forms of personal writing in the first semester and then "graduating" to the more impersonal argumentative writing in the second.[1] This progression is homologous to a move from the working classes, associated with the personal or subjective, to the middle and upper classes, associated with the impersonal or objective. The depersonalization of discourse demonstrates one's distance from things, which is to say, distance from necessity—enabling, Bourdieu (1984) argues, the aesthetic disposition (53-54). One becomes distanced from one's self—i.e., nothing is taken "personally." By taking an impersonal stance, one is more capable of appealing to logic rather than emotion—and logic is what higher education is presumably about (Fish 2008).

Although sometimes introduced in the first semester writing courses, the controversial issue or position paper typically dominates second semester required writing courses, for which reason both the ACT and SAT have made it the model for their timed writing assessments, claiming it as the central kind of academic discourse (ACT 2003,Table 1.8; College Board 2004). The genre is reality-focused insofar as the writer is not supposed to distort information for persuasive purposes; on the other hand, the writer should carefully consider readers' responses, arranging the information and arguments, such as putting the most significant reasons last or delaying the position statement if the writer anticipates readers who would initially be offended by the writer's beliefs. The prototypical controversial issue essay, with its roots in

---

1. I looked at the Web pages of 114 universities for whom composition requirements were readily available. Seventy-four of these were non-liberal arts schools. Of these, 61, or 82%, made argument the focus of their second writing course. With 43 of these, argument was the clearly stated focus. Others described their courses in terms like writing about literature, thinking critically and backing up your arguments, or clearly stating and defending a controversial thesis; from descriptions like these, I inferred argument as the central genre. The notable exceptions were courses that emphasized discipline-based writing. Also, courses that were portfolio-based tended to be more catholic in the genres in which they had their students write.

Sheridan Baker's *The Practical Stylist* (1962), usually focuses on some issue with a thesis that is worth writing about only if the writer can imagine an antithesis that some readers would support. The essay is framed by an introduction with a specific thesis at the bottom of a triangle, a paragraph or two acknowledging oppositional points of view with rebuttal and several paragraphs presenting the writer's reasons for promoting the thesis, with each claim being supported by evidence and logic linking the evidence to the claim. The writer then restates her thesis in the conclusion and broadens out to implications or the need for further investigations. The formulaic nature of this genre, only slightly more diffused than the five-paragraph essay, is most likely the reason behind its popularity. It gives teachers and young writers something to hold on to.

Although the formulaic nature of this genre might appeal to students with working-class origins, several central features of this argumentative genre conflict with the working-class ethos. These are objectivity, multiple perspective, explicit language, stance, and dialogism. I will explore in this chapter the class biases of these features and conclude with an analysis of working-class resistance to argument, based on a misinterpretation of the genre.

Objectivity

The relationship between social class membership and objectivity is as old as Plato's parable of the cave. According to Plato's parable, the working-classes—or any colonized social group (women and slaves, in particular)—are dominated by subjectivity. They are chained to their points of view from which they can see only the stories cast on the wall. The middle classes are more likely to understand the shadows are cast by actors because members of these classes have access to the scene of manipulation. Members of these classes turn their eyes from the wall to the ledge behind them where the scenes are being staged. The higher social classes have found their way out of the cave to where they are able to see the things in themselves, the modernist *ding an sich*. These are of course the scientists and New Critics. But the truly elite social classes, dramatizing Bourdieu's (1984) thesis on distancing, are not beguiled by things, that is, by the particulars; instead they cast their eyes to the heavens, where they see the Forms. This is the paradigm of a social class hierarchy, at the upper reaches of which form is everything. Reference to the self and one's concerns—e.g., through expressive discourse—are in poor taste, a betrayal of one's lower-class heritage (Bourdieu 190-200).

The objectivity associated with the higher social classes is key to the argumentative genres that Kinneavy (1969) classifies as referential discourse with an obligation to render external reality faithfully. The best essays in these genres appeal to reasoned discourse with a balanced tone. When writers slip into pathos as an appeal, academics will be irritated by the writer's attempt to draw the reader into the emotional game. Emotional discourse, with its origins in epideictic rhetoric, may work well in political rallies but not in academic settings. The pose of objectivity, in fact, lay behind the injunction against use of the first person pronoun through the first half of the 20th century. Although the first person is now admitted in most writing classes at the college level, it must still be an objectified "I," the writer in possession of her reason, language, and stance. If Bourdieu's thesis on the relationship between social class and the distancing effect is accurate, then one's attitude toward objectivity corresponds to the social class level of the writer. Students from the higher social levels will more readily accept the objective pose because it is more a part of their ethos and language codes than it is for working-class writers.

In an analysis of his own upper-class ethos, Nelson Aldrich (1996) calls sublimation the key "operating principle of the upper-class aesthetic" (214). The upper-classes are socialized in their primary culture into sublimating their emotions, desires, and needs because they are born into the condition of being *above* it all. In their desires to be like the upper-classes, members of the middle classes pretend to the sublimation naturalized in the upper-classes. For middle-class members, the struggle for sublimation marks their primary Discourse—it is what they are born into, the *pretense* of distance through an objectified "I" (Bourdieu 1984, 253).

Academics have adopted an unnaturally naturalized tone to their objectified discourse. It's a superficially distanced tone that may seem natural to those who have been raised in the parlance, but to a working-class sensibility, the tone rings false. Here is Lynn Bloom's (1996) paradigmatic middle-class discourse:

> Most of the time, the middle-class orientation of freshman composition is for the better, as we would hope in a country where 85% of the people—all but the super-rich and the very poor—identify themselves as middle class (Allen). For freshman composition, in philosophy and pedagogy, reinforces the values and virtues embodied not only in the very existence of America's

> vast middle class, but in its general well-being—read promotion of the ability to think critically. (655)

Bloom's prose is a model for the tone rewarded by college teachers. Although Bloom is close to her subject here (social class and writing instruction), her tone is balanced and cool. One simply has to parse the first independent clause to see this:

> Most of the time
>
> the middle-class orientation
>
> of freshman composition
>
> is for the better

Her prose reads smoothly because of this control and balance—in this case, we have four lines with two strong stresses each, the second and the third lines having third secondary stresses (tá tion and sí tion) that create a rhyme. The rest of the sentence unrolls with the same balance and control to end with two dimeters

> identify themselves
>
> as middle-class

repeating her initial noun and subject with the final stress directly on the word she is emphasizing—"class." Throughout these two sentences (which I pulled almost at random from her writing—my eye having been caught by her reference to critical thinking), she uses parallel elements:

> the super-rich and very poor
>
> in philosophy and pedagogy
>
> the values and virtues
>
> not only in the very existence . . . but in its general well-being

She uses formal language when she has a choice: ("for" instead of "because") and the self-conscious appositive "—read promotion." This is good writing, the kind that gives academic readers confidence in the writer. But from a working-class sensibility, that appearance of control (like being carefully dressed) makes the writing and writer suspect, or as Lindquist (1999) puts it in her analysis of the barroom regulars, "they hold in suspicion those performers who are obviously adept at the game—the better one speaks, in other words, the less he or she can be trusted" (236). Working-class writers write ragged. Their words spew out, an eruption of thought and emotion, which is perhaps, as Ong argues, the consequence of being raised in a primarily oral culture, of not being used to the objectification of the self as text, of being "out there" where one can be re-seen and revised.

While reviewing a textbook proposal a few years ago, I came across a striking example of student writing in the distanced style—perhaps the kind of prose Lynn Bloom may have written when she was in high school, prose that wears a tightly buttoned, Peter Pan collar. Here's the introduction to that student essay:

> In his poem "Sonnet," Percy Bysshe Shelley introduces us to a bleak world that exists behind veils and shadows. We see that although fear and hope both exist, truth is dishearteningly absent. This absence of truth is exactly what Shelley chooses to address as he uses metaphors of grim distortion and radiant incandescence to expose the counterfeit nature of our society.

The writer is clearly distanced from her text. We don't see her in it at all—in fact, one gets the feeling of the writer overworking each line to get the balanced tone and ornate style we saw in Bloom's writing: "bleak world that exists behind veils and shadows"; "although fear and hope both exist, truth is dishearteningly absent"; "metaphors of grim distortion and radiant incandescence to expose the counterfeit nature of our society." The writer self-consciously steps back to choose words like "dishearteningly" and parallel phrases like "grim distortion" and "radiant incandescence." These are prose words, words that appear only in an overly written piece of writing. They are certainly not words the writer would speak, that is, if she valued her friends. This prose is also the kind her teachers probably taught her to write and that will gain points in her liberal arts courses, which is why the author of this rhetoric, whom I very much admire, chose to use it as an example of strong student writing.

Readers with different social class trajectories might read this style differently. College teachers who were either born middle-class or worked hard in school to achieve middle-class status may like this kind of writing, but working-class academics—at least those who have not rejected their working-class backgrounds—may be more inclined to dislike it, or at least respond ambiguously to it. On the one hand, the academic part of me says, if this style will help writers fight their ways through the academy, then I'll teach it. The working-class side of me, however, shrinks from this kind of overly self-conscious prose, particularly when I find myself writing it.

Although I resist the necessity of teaching my working-class students how to distance themselves from their texts, the ability to objectify oneself (or one's text) is necessary both for a better understanding of one's place in the world and for improving one's writing skills. We need to understand, however, that because their primary Discourses may be in conflict with this pose of objectivity, working-class students may be inclined to resist it. We might attribute their resistance to their uneducatability but we could more accurately understand it as a contradiction between the essential features of their primary and secondary Discourses.

Multiple Perspectives

Working-class students may also be less likely than middle-class students to see events from multiple perspectives. Being able to see events from multiple perspectives is in part the consequence of being able to step back from one's subjectivity. This step-back is enabled by two socioeconomic factors. The first follows from Bourdieu's (1984) thesis of elites having the leisure to step back. Bourdieu plays with the "ease" in leisure—the ability to feel at home in one's situation, which Bourdieu equates with the elite's ownership of what counts as culture (2, 255-56).

The second factor links to Bernstein's (1971) thesis on the social class differences in role-playing and social experiences. Bourdieu, Bernstein, and Lareau (2003) document the relative closed-role system in the working classes as opposed to the open-role of the middle classes. In essence, the higher one's social class, the more varied the roles one is required to play, simply because members of the higher social classes find themselves in more varied social circumstances than members of the working classes do.[2] As well as being exposed to varied social situations in their everyday lives, higher social class members travel more, are exposed to

2. As in almost all generalizations about social class hierarchies, they bend back on themselves when they approach the extremes. The elite, or the upper one percent of our social system do not mix, very much as lower working class do not mix. It is no accident that their habitus are closer to each other than to the middle classes, for instance, the different emphasis on seeming and being. Both the elite and working-classes emphasize being. The middle-classes, intent on moving upward, emphasize seeming at the expense of being.

different languages, different environments, and different modes of being. In Lareau's study, the contrasting travel experiences between the fourth grade, working- and middle-class children couldn't have been more explicit: the urban working-class kids rarely travel outside the safe zones of their neighborhoods; the middle-class kids travel all over the city on a weekly basis, out-of-town, out-of-state, and occasionally to different countries—travel habits that are frequently modeled by their professional parents, some of whom fly to other cities two or three times a week in their normal course of business. Lareau cites Garrett whose father averages three nights a week out of town. Garrett's own participation in sports and other competitive activities frequently takes him out-of-state for various tournaments (42-44).

Elements of their habitus such as travel habits and multiple social positions naturalize for higher social class members the condition of being able to see from many different points of view, a central feature of academic argument. In my introduction to this chapter, I noted that firstyear students' experiences are circumscribed, but the degree of circumscription is a function of one's social class origin.

The postmodern assumption of evanescent subject positions presents an additional unacceptable framework for the working-class student. Because of their circumscribed experiences, working-class students tend to resist the notion of shifting identities—for them, changing who you are to respond to the social context is what middle-class people do (Bourdieu 1984, 252-53; Seitz 2004, 55). For the working-class person raised in circumscribed environments, identity is fixed. For the middle-class students, whose parents have to learn how to respond to varied rhetorical situations, identity and meaning are contextualized. This latter notion of shifting, contextualized perspectives is what we reward in student writing.

Dialogism

Dialogism is the bedrock of academic argumentation. Not only does the writer have to imagine alternative perspectives, but she also has to be able in her discourse to speak to representatives of these different perspectives. The writer has to construct the appearance of presenting oppositional perspectives fairly and answering their arguments with more persuasive arguments of her own.

This dialogic strategy, emphasized in Rogerian rhetoric, is compatible with the middle-class habitus in which argument as reasoned discourse is encouraged, a consequence of parents who are constantly negotiating with others in the workplace and who bring this way of seeing the world home with them. Parents who are generally in non-negotiating situations in the workplace internalize a monologic way of seeing the world and reproduce it in their homes. At the dinner table, middle-class parents might attempt to engage their children in dialogic conversations in which ideas, politics, and public policies are discussed, listening carefully to their children's positions and encouraging them to make their ideas explicit and explore alternative possibilities. Lareau (2003) writes that Alex's middle-class parents frequently work to develop his political awareness by discussing political events with him at dinner (119), a highly unlikely dinner table conversation in a working-class home, where, as Mueller (1973) remarks, "children are to be seen and not heard" (59).

The correlation between dialogism and social class is documented in Anyon (1989), Bernstein (1971), Bourdieu (1974), Brice Heath (1983), Lareau (2003), and Mueller (1973). Brice Heath documented the differences between working-class and middle-class parental habits while reading stories to young children: the middle-class parents encouraged dialogue with the children and, in effect, with the text, as opposed to the working-class parents who taught their children how to sit and listen, often far beyond the point at which they had any interest in the story (225-26). Brice Heath claims the parents were replicating in their home their different interpretations of their own schooling experiences. If Anyon's research is indicative of a general trend, the parents in Brice Heath's study are in fact correctly interpreting what their children need to know—at the lower grades, at least. Anyon's research, although limited in scope, indicated that the different social class home orientations are reproduced in the primary schooling years. The working-class schools are role-driven, authoritarian, complete with bells. As the social class membership of the other four schools was higher, so was the amount of freedom, participation, and student self-direction. In schools dominated by children of the executive elites, there were in fact no bells.[3] The dialogic nature of the executive elite was exemplified in the math class in which there was no notion of "right" answers. The supposedly "right" answers could always be challenged, which the teacher encouraged the students to do. In the following classroom dialogue while reviewing answers, the teacher said,

3. My wife is a librarian in the most exclusive private school in Baton Rouge. No bells.

> "Raise your hand if you do not agree." A child says, "I don't agree with 64." The teacher responds, "OK. There's a question about 64 (to class). Please check it. Owen, they're disagreeing with you. Kristen, they're checking yours." The teacher emphasized this tactic repeatedly during September and October with statements like, "Don't be afraid to say if you disagree. In the last (math) class, somebody disagreed, and they were right. Before you disagree, check yours, and if you still think we're wrong, then we'll check it out." (Anyon 1989, 84)

Soon, the students weren't talking in terms of right answers but in terms of whether they agreed. This pedagogical strategy encourages a dialogic attitude toward school, authority, and society in contrast to the working-class school in which rote learning was the rule. The teachers in working-class math classes taught the children how to follow certain textbook-dictated steps to solve problems. Their work was evaluated on the basis of how well they had followed the steps. One girl in the working-class school, for example, said in response to one problem "that she had a faster way to do it and the teacher said, 'No, you don't; you don't even know what I'm making yet. Do it this way, or it's wrong' (Anyon 1989, 74).

The working-class children learn in their homes and schools that the world is a fixed order (Freire [1970] 1995, 73) to which they have to accommodate themselves, muttering, perhaps, under their breath; but the middle-class kids learn that reality is malleable, that people in their condition can in fact effect change by speaking to the world, which in turn speaks back. Their parents speak to political representatives, to doctors, to lawyers, to store owners and expect to be heard. The higher your social class, the more you are heard (Bourdieu 1975, 405-406). It might well be the case that the more you are heard, the more you hear.

Explicit Language

The relationship between one's language and one's social and working environments includes not only grammar, words, and syntax but also kinds of words, use of modifiers, specificity, tone, rhythm, rhyme, structure or genre of utterance, and assumed knowledge constituting gaps within the text. Bernstein (1971) aligns these differences along an implicit/explicit continuum with the working classes oriented toward the less privileged member—the implicit. Explicit language depends on having a large vocabulary allowing one to make distinctions and the syntactic versatility to make clear the relationships between linguistic elements. By privileging explicit language, we are unsurprisingly undermining the probability of scholastic success for working-class children, who grow up in homes with fewer linguistic resources and little need to make themselves understood by strangers.

Research has shown that linguistic experiences are significantly determined by our social class membership. At a basic level, Hart and Risley (1995, 32) documented the average number of words spoken each day to children at ages 13–36 months in the classes: professional (2,153 words), working (1,251 words), and welfare (616 words). By age three, the average child from a professional-class family had twice the vocabulary of a child from a welfare-class family (164-165). If this relationship between social class origin and language development continues, we can speculate that by college level, the student from a professional-class family has a much richer linguistic resource to draw on than a welfare-class student to make his or her meaning explicit.

Moffett (1968) theorized that language development is dialogic, a process of internalizing others' speeches and making them one's own (see Bahktin 1981; Vgotsky 1962). Moffett explains that an utterance by an interlocutor can become internalized and replayed in the child's speech as an embedded phrase or clause (72-83). On a more abstract level, dialogues are internalized and replayed as monologues within new dialogues. In essence, Moffett claims that our current linguistic resources are a consequence of previous linguistic experiences, our monologues the result of previous dialogues.

Lareau (2003) convincingly documents the degree of the differences in linguistic resources among middle-class and working-class families. Ms. McAllister, a working-class mother, uses language in a predominately functional capacity. She sees no need, Lareau points out, to spend time trying to enrich her son Harold's language (139). When the adults and children in the McAllister home were together in a room, Lareau reports, "Short remarks punctuate comfortable silences. Sometimes speech is bypassed altogether in favor of body language—nods, smiles, and eye contact. Ms. McAllister typically is brief and direct in her own remarks" (146). Her children are told to do things, frequently with one word directives. She "sends the children to wash up by pointing to a child, saying, 'Bathroom' and handing him or her a washcloth" (147).

We see this abbreviated, working-class discourse again when Harold and his father are shopping to prepare Harold for church camp. "Harold picks up a plain blue [beach towel] in the bottom rack. He holds it up. His dad says, 'You want a plain one?'" Harold nods. This one-way conversation is followed by an exchange in which Harold's father tries to get Harold to accept a good buy on a peach color towel set, to which Harold "shakes his head. 'Them girl colors.'" In the subsequent exchange, the father utters short sentences, to which Harold either nods or shakes his head. The ethnographer reported that through the entire exchange Harold used fewer than ten words (Lareau 2003, 148).

In contrast with the brief directives and responses observed in the working-class families, Lareau (2003) writes that the middle-class families she and her researchers observed rarely issued directives or corrected disobedience with physical threats, the common modes of socialization in the working-class families.

> Instead we observed them repeatedly, systematically, and determinedly use verbal negotiation to guide Alex through the challenges in his life. As Basil Bernstein has noted, rather than using authority based on positions (e.g., that of being a parent) middle-class parents prefer negotiating interactions with their children in a more personalistic fashion. They use reasoning to bring about a desired action, and they often explain *why* they are asking children to do something. (116)

Lareau (2003) uses as an example a dinner table conversation in which Alex's father is trying to argue his son into eating string beans, he says,

> "How are you going to beat up Fritz if you don't eat your vegetables?" Alex shook his head as he picked up a string bean with his fork, "I am not going to fight him!" Terry, smirking: "Are you going to let him bully you like he does the other kids." Alex, alternating stares between his father and his plate: "I'll fight him if I have no other choice, but I'll tell one of the teachers so he can get suspended." (118)

Not only is Alex allowed to counter his father's argument, his logic is supported by his mother who says, "That's right, baby. You do not have to fight. There are better ways to resolve conflict" (118).

This kind of negotiation and acknowledgment of the child's superior logic (at least in the mother's eyes) would rarely be tolerated in a working-class family. The working-class parent would far more likely resort to position-authorized authority, the logic being supported by the parent's interpretation of his or her responsibility to punish disobedience. When Ms. McAllister is faced with similar resistance at the dinner table, she resorts to argument through volume. The ethnographer reports: "'Mom yells (loudly) at him to eat: 'EAT! FINISH THE SPINACH!'" (147). In my home, the culinary etiquette was the same: we ate what was on our plate, or else . . . (as Bill Cosby [1968] memorably put it in "To Russell, My Brother, Whom I Slept With") THE BELT![4]

The effect of these different modes of socializing children is dramatically asymmetrical language development. Lareau (2003) says that through the consistent negotiation of behavior, middle-class children like Alex developed extensive, adult vocabularies. When his father chides him for changing his mind about his favorite cars, Alex can reply, "This is America. It's my prerogative to change my mind if I want to." Alex's parents are constantly inserting words like prerogative into their conversations with the implication that Alex should take them up in his equally constant negotiations and arguments with them (130), a middle-class pattern that accounts for Moffett's (1968) observation from his research that "white middle-class fourth-graders [were able to] write rings around the ninth-grade ghetto children in sensory and memory writing" (34-35).

In brief, the working-class children are trained to communicate with gestures, using as few words as possible, emphasizing an economy of expression. The middle-class children are taught to use sophisticated words. They learn to play with language and take pleasure in that play, exploring double entendres (Lareau 2003, 45) and irony (119), both markers of higher social class membership (Bourdieu 1984, 34). If language is play to the middle-classes, it is work to the working classes, and the less of it, the better.

Stance

In their "Preface" to one of the most popular books on argument, Bartholomae and Petrosky (1993) recommend the following to students: "To take command of complex materials like the essays and

4. I am conscious of the values associated with the differences between social classes and modes of parental control. The researchers observed (and Billy's mother reported) the use of "the belt" on the average of once a week (228-229). For various reasons, physical punishment or threats of physical punishment as a form of parental control is a function of social class membership. The higher one's social class, the less the use of physical punishment. Now that I am a middle-class professor married into the intellectual fraction of a middle-class family, I disparage parents hitting their children. In fact, the sight of a working-class mother yanking or slapping her child in the grocery store upsets me. This reaction is situated.

stories in this book, you need not subordinate yourself to experts . . ." (10); and again, "It is up to you to treat authors as equals. . . . [a] reader takes charge of a text" (11). Legions of composition teachers have taken up this heady directive to teach their students to read and write against the grain without considering how this advice privileges middle-class children, who are trained to assume precisely this stance.

Reading against the grain of one's parents would be considered "talking back" in the working-class homes Lareau (2003) studied; but the middle-class parents clearly understood that by training their children to challenge them, they were preparing them for the academic and professional worlds in which "talking back" is a sign of competence. When Alex's father tells him that if he copies a riddle to complete his riddle assignment, "Someone can sue you for plagiarism. Did you know that?" Alex responds as few working-class children would ever dare do, "That's only if it is copyrighted" (119). Rather than feel challenged, Mr. Williams, a lawyer, may have felt he had a young lawyer on his hands, an interlocutor who was able to spot a broad generalization and challenge the claim by noting that the generalization applied only to a limited set of situations—and that the case in point did not fall within that range. It is worth noting that Alex, a fourth grader, has made "copyright" his own word (Bakhtin 1981, 293), a word that Harold, a working-class child, had probably not heard.

Alex's challenge to his father, Lareau reports, is far from uncommon in the Williams family—in fact, the parents seemed to encourage Alex to speak out whenever he disagreed with something either of them said. Lareau's researacher reported, "They seemed delighted with his overall development and they were unperturbed when he periodically used the skills they had taught him to challenge their authority. For them, the benefits of 'developing' Alexander outweighed the costs" (129). The benefits lay in the future—they know Alex with his sense of entitlement will be professionally successful. In the upper-middle class themselves, they know the world for which they are training Alex.

Lareau (2003) gives a striking example of another middle-class parent modeling an empowered ethos after her daughter had had an emotionally upsetting class with her gymnastics teacher. Rather than respond by telling her daughter to work harder, as working-class mothers would do (see Brice Heath 1997, 348), Ms. Marshall confronts the teacher about the teacher's overcorrecting behavior. When the teacher says the daughter just has to learn more of the terminology, Ms. Marshall replies, "Look, maybe it's not all the student." After she gets home, Ms. Marshall further demonstrates her empowerment—with her daughter watching, she calls the owner of the gymnastics school and convinces him to start another class of advanced beginners in response to her daughter's inability to cope with the demands of the teacher in the intermediate class (171). The mother is clearly teaching through her example that her daughter should not accept situations—if she is dissatisfied with them, she should take steps to change them, confident that she has the authority necessary to intervene and negotiate change.

Similarly, Alex's mother uses a visit to the doctor's office as an opportunity to train Alex to interact with the doctor from an empowered position. Ms. Williams tells Alex in the car, "Alexander, you should be thinking of questions you want to ask the doctor. You can ask him anything you want. Don't be shy. You can ask anything" (Lareau 2003, 124). As a consequence of Ms. Williams' preparation, Alex readily engages with the doctor, asking questions about the doctor's interpretation of Alex's height, interrupting, and correcting the doctor when he says that Alex is 10 (Alex is only nine [Lareau 2003, 124]). When the doctor asks whether Alex has any other concerns, Alex freely explains that he is worried about some bumps in his armpits, assuming control of the direction of the conversation. Lareau notes that the transition of the conversation's direction "goes smoothly. Alex is used to being treated with respect. He is seen as special and as a person worthy of adult attention and interest. . . . He is behaving much as he does with his parents—he reasons, negotiates, and jokes with equal ease" (126).

In contrast, working-class parents train their children *not* to take "charge of a text" (Bartholomae and Petrosky 1993, 11), that is, of a rhetorical situation in which the working-class person feels he or she is expected to defer to authority, reproducing the parent's rhetorical situation in the workplace. Harold, a fourth-grader like Alex, learns this deference through the example of his working-class mother's behavior. Although Ms. McAllister is described as an athletic woman who in the home has "a highly developed sense of humor and a booming voice," she turns into another person in the doctor's office. Lareau (2003) describes her as "quiet, sometimes to the point of being inaudible." When the doctor tries to make eye contact with Ms. McAllister, she looks either at the piece of paper in her hand or down to the floor and mumbles monosyllabic answers like "Yeah." Harold is watching—and learning. Like his mother, he answers the doctor's questions with monosyllables and in a low voice (157), most unlike Alex who has been trained to treat the doctor as an equal.

When Harold's mother meets with his teachers, Lareau's (2003) researcher reported that "[t]he gregarious and outgoing nature she displays at home is hidden in this setting. She sits hunched over in the chair and she keeps her jacket zipped up." Ms. McAllister is surprised when the teacher tells her that Harold has not been turning in his homework. Ms. McAllister says he has done it at home, but "she does not follow up with the teacher or attempt to intervene on Harold's behalf" (157). Echoing Brice Heath's (1983) remarks on the working-class whites in her research, who "turn over to the school their child's education" (148), Lareau says that Ms. McAllister felt that "it is up to the teachers to manage her son's education. That is their job, not hers. Thus, when the children complain about a teacher, she does not ask for details" (157)—as if the children have been shipped to an alien world where they have to learn to sail on their own. This disassociation is an effect of social class alienation: as Marx argued the working-classes were alienated through industrialization from their labor, so are the working-class parents alienated from their children's educational processes. Education is closed off in urban working-class schools as if by a steel fence. The fence stands like Ellis Island as the sign of a world isolated from the working-class community in which it is placed. The school is a transition point, a moment, as Bloom (1996) trenchantly claimed about freshman English, of disinfection. To the working-class parent, the teacher is the middle-class representative in control of that "other" world—and of the child after she passes through the fence or disinfecting pool.

If we assume that the logic governing the relationships between social classes governs the interpretations of one's own power to effect change, we can see how Lareau's (2003) research predicts the problems working-class students have when they walk into college writing classes in which the power to intervene is consistently privileged over the working-class ethos of skin-level accommodation. Lareau claims that the middle-class children in her study learned their sense of place, their right to protest and correct. The working-class children learned to be silent, reserving their protest for the sanctity of the home where they might bitterly complain about teachers (163) as working-class parents complain about their bosses in the home—or in the bar.

The Place of Argument

One reader of this manuscript challenged my claim that working-class children are not trained to argue in their homes and working-class schools. He cited Julie Lindquist's (1999, 2002) auto-ethnography of barroom "working-class rhetoric" (2002, 3) as evidence of the place of argument in working-class discourse. Lindquist interprets barroom arguments as an index of the mechanisms through which working-class people establish their identities (vi, 16-17, 41). She analyzes a plethora of instances in which the Smokehouse regulars bait each other—and Lindquist, a bartender—into social and political arguments. In contrast with the implications of Lareau's (2003) and Bourdieu's (1984) research (see 405-406, 427), Lindquist claims that political arguments, in particular, are central to working-class identity (10).

I take Lindquist's research seriously. Although I didn't hang around working-class bars as an adult, much less as an ethnographer, I spent more time than I should have in my late teens in them. Even through the refracting lens of time and the different perspective of a teenage participant, I recognize the regulars, the social structures, and the barroom banter Lindquist documents in her auto-ethnography. I question, however, Lindquist's claim that the barroom is a valid microcosm of working-class culture. As Lindquist points out, the barroom culture is predominately male with a few of the men's partners and some of the female employees, Lindquist among them, admitted to the group of regulars; one would be hard-pressed, therefore, to see the barroom culture as a fair representation of working-class culture as a whole, which also includes women, children, and a fair portion of both men and women who would bristle at the notion that the group of men who habitually hang around the bar after work represent their culture. We should also note that this is an urban, working-class bar, a space, as Lindquist describes it, where the working men can gather to create powerful social identities, perhaps to counter their underling identities in their work situations (see Freire [1970] 1995, 48), re-establishing their self-authority through drinking.

In answer to my reader, I am reframing Lindquist's research as a special rather than a general case. As a special case, barroom discourse may very well reveal a suppressed imperative of working-class culture, but we need to acknowledge that this discourse represents only a part, and quite possibly a minority, of working-class adults. Literature frequently normalizes the bar as a site of the working-class male, but it's quite possible that this narrative of the working-class bar, usually written by middle-class observers or cross-overs like D. H. Lawrence is another marginalizing strategy told by the outside culture. When viewed against the working-class homes Lareau documents, the Smokehouse (Lindquist

1999, 2002) seems like another country. My reader's caution against over-generalizing is, however, well-taken. I acknowledge that people in working-class culture argue, but there is a notable difference in the rhetorical situations in which children from the different social classes learn to argue.

In Stephen Garger's (1995) account of his struggle to adapt to academic notions of argument, he writes that any brief arguments with his parents would end with "because I said so. . . . My parents never listened," he says, "and after a while, neither did I" (49), a lesson that is the antithesis of what academics need to do. Garger also associated argument with the kinds of arguments in which he and his friends engaged on the street, generally over who had done what in pick-up basketball, punctuated by shouts, name-calling, swearing, and face-offs that led either to fights or "do-overs." He and his friends didn't argue by citing evidence and linking their evidence to their claims; rather, "might made right" and "the only response to a fellow ignoring or contradicting an explanation three times is to yell and go for the throat" (49).

The might-makes-right logic of argument in the working-class home and on the street may account for cross-overs' rejection of argument in the classroom, a consequence of their inabilities to distinguish a difference in kind. In a presentation at the Conference on College Composition and Communication in 2000, Nancy Mack said, for instance,

> I hate argument. My mother and father, before me, hated arguments. I guess that it runs in the family. In our working-class home when I was growing up, to be accused of wanting to argue about something was worse than cursing. It was considered an abusive practice used to brow-beat someone until you got your way.

Alfred Lubrano (2004), a journalist who wrote a book on the sound barrier through which working-class kids have to pass, summarizes the relationship between argument and social class like this: "In blue-collar homes, there's rarely such a thing as a civil argument. Working-class people have two speeds: silence and rage. It's the middle class that debates things, able to conduct an argument without becoming emotional; working-class people yell" (65).

By linking argument with shouting and the threat of violence, working-class students and academics tend to handicap themselves by meeting argumentative situations with silence, which is precisely how Garger met a challenge to his claim in a faculty discussion that he was leading.

Rather than listen carefully to the faculty member who was questioning his claims and answering the challenge with reasons supporting his position, he clammed up and went into an inner boil, furious at the faculty member who had questioned Garger's assumptions. His intractability marked him as someone who did not yet belong.

Garger recounts how one of his friends helped him see argument from the middle-class, academic perspective. While they were walking to another faculty meeting,

> [William] said he was looking forward to the gathering because there were sure to be some good arguments this time. I was surprised and asked, "You're actually looking forward to arguing?" His reply surprised me even more: "I like arguing. It's a good way to compete intellectually and have fun. The issue doesn't have to matter." He held the door open for me as we entered the meeting room, gave me one of his rare smiles and said, "Academics like to argue." (50)

William was opening the door to membership in this more privileged community in which argument is game, a game middle-class children are trained to play as much as working-class children are taught to avoid it.

After his conversation with William, Garger analyzed argument as a discourse form distinguishing his home community from the one he was joining. He did not treat one form as deficient and the other as competent; they were merely different, distinguished by the social groups that used them. In their arguments over theories, middle-class professors, Gargar points out, were doing what his blue-collar playground buddies were doing when they got into shouting matches over the value of different players on professional baseball teams—creating a space within which the speakers were able to show off their knowledge. The real difference lies in which discourse form is valued by the people in power. The people in power argue like William; the people out of power argue like the people in Garger's home community: they shout, replacing logic with volume—and, significantly, only among peers, never against people in superior positions of authority, reinforcing Bernstein's (1971) distinction between the working-class position and the middle-class person ethos. The working-classes are trained to grant authority on the basis of the social position of the person, while the middle-classes are trained to grant authority on the basis of the arguments they make.

Working-class kids certainly do learn how to argue—but in a mode and rhetorical situation that is at odds with the kind of argument we teach in our writing courses. Working-class argument is governed by volume, invectives, and physical gestures; or as Bernstein (1971) would put it, by implicit rather than explicit language. In addition, it is practiced—and to me this is the salient point—among institutionally sanctioned peers. Kids can argue with kids *outside*, but they cannot argue with adults or in the home (see Lareau 2003, 142, 154). Kids can argue *only* in the space marked off for kids only. I interpret a corresponding logic underwriting Lindquist's claims. The bar is the place marked off from the workplace, marked off from the structures that govern social relationships determining who gets to say what when it counts. In the bar, you drink, and the alcohol puts you outside the *normal* discourse relationships. You can make your point by shouting, as the kids on Garger's playground made theirs. The bar, that is, is a playground, a carnival (Bakhtin 1981, 23-25), in which the normal discourse and social conventions are turned upside down. But the sun comes up, the streets are swept, and a little bit later, everyone goes back to school.

6

CULTURAL STUDIES AND COMPOSITION

Teachers in the social strand of the critical thinking and writing have attempted to fill the putatively empty rhetorical situation of required writing classes by making culture the object of study. The trajectory of this strand moved from the early writing-about-literature phase through writing-about-self to writing-about culture.[1] The first phase was a natural outgrowth of the field's origin within literature departments—the graduate students teaching the course assumed that "writing" meant writing about literature, as they had been trained to do. "Real" writing was the writing they were writing about—and the people who wrote "real" writing were literary figures. So they imagined their task, as many writing teachers still do, as one of helping their students become "writers"—or at least writers who write about writers, biding their time, perhaps, until their own novels would sell. The missionary and inevitably disappointed zeal with which the writing-about-literature graduate students approached their task led in the early 70s through the mid 80s to a writing-about-self pedagogy, inherited by a misreading of James Britton et al.'s (1978) *The Development of Writing Abilities (11-18).*[2]

The expressivist tradition focused on student writers learning how to get in touch with themselves as writers, turning inward the belletristic tradition out of which rhetoric and composition grew. Students first learned how to write for themselves and then moved outward to write for others, a model popularized by Britton et al. (1978) and James Moffett (1968) in *Teaching the Universe of Discourse.* The personal essay in the expressivist tradition was the privileged genre at the postsecondary

1. And more lately, writing about writing—a phase that is outside the scope of this book (Downs and Wardle 2007; Miles et al. 2008; Bird, Downs, and Wardle 2008).

2. Both Britton et al. and Moffett promoted instruction in the full range of genres. In Britton et al.'s construct, expressive writing is only one of four supra-genres. The others are transactional, poetic, and "additional" (primarily different categories of school writing).

level. In *Critical Thinking and Writing*, Thomas Newkirk (1989) promoted the personal essay as an answer to the perennial problem of disembodied writing in required writing classes (Britton et al. [1978] called this kind of writing "perfunctory," 8; Ken Macrorie [(1968) 1976] called it "Engfish," 4); Newkirk claimed that the personal essay was also the necessary link to teaching students how to think critically. According to this tradition, writing is a way of seeing anew (S is a CT in X if S can do X by imagining ~E or ~any subset of E), of learning more about oneself and one's culture—a way of getting in touch with deeper and generally obscured truths. In the personal essay, the writer invites readers to see the inner workings of the writer's mind as it wanders in a Montaignesque fashion to a *re*vision of whatever subject the writer had set out to explore.

In 1993, Kurt Spellmeyer published *Common Ground*, a thoughtful exploration of the purposes of education and the personal essay's function in achieving them. Both Newkirk's and Spellmeyer's interpretations of our purposes in teaching writing draw on the Arnoldian "legacy of liberal culture" (Berlin 2003, 35). As Berlin describes it, liberal culture is rooted in an elitist tradition of training upper crust students in the liberal arts so that they will know how to govern by inheriting "the best that is known and thought in the world" (Arnold 1913). Newkirk and Spellmeyer democratize the Brahminical tradition, spreading liberal thought outward through required writing programs. Spellmeyer wants writing teachers to encourage students to push against the edges, abandon the security of what has been known and plunge like Lord Jim into the darkness. Exploring the limits of thought about self and culture seems for Spellmeyer to be the dominant purpose of writing. Drawing heavily on Foucault's "Discourse on Language," Spellmeyer positions his argument within an Institution/Intuition dichotomy. Institution represents rules; Intuition represents impulse, the unregulated. Framed within Bakhtin's theory of language and culture, Institution is aligned with the centripetal force giving utterances shared meanings, and Intuition is aligned with the centrifugal force pulling old language apart, creating a space for new meaning. Language is of course a synecdoche for culture, the pull of tradition against the force of revolution (Peckham 1997).

Spellmeyer (1993) dismisses as shortsighted instrumental motives serving commerce and industry, in which clear communication is imperative. Spellmeyer links clarity to writers' knowing what they want to say before they say it as opposed to the possibly muddled adventure of figuring out what you want to say *through* saying it, the safety of tradition against the risk of innovation, the anchor of the conservative against the mainsail of the radical. Although explored primarily along epistemological lines, Spellmeyer's thesis clearly spills into the social class and political domains.

The critical pedagogy associated with Berlin has gained more traction than Newkirk's and Spellmeyer's promotion of the personal essay. Although all three have the end purpose of cultural critique, Berlin's 1987 attack on the expressivist tradition (145-54), with which the personal essay is linked, undermined Newkirk's and Spellmeyer's pedagogies. Berlin (1988) interpreted the expressivist tradition as the result of a naïve, romantic individualism that inadvertently plays into the hands of the late capitalists (487). In place of the expressivist and cognitivist traditions, Berlin argued for a social epistemic rhetoric with its roots in the Frankfurt school. Responding to the easily anticipated charge that they are attempting to indoctrinate students into leftist thinking, theorists aligned with the social epistemic tradition consistently claim they want students only to think critically. The assumption, inherited from Freire, is that when students think critically about late capitalist culture, they will perceive the social injustice embedded within it and be moved to do something about it. The social-epistemic tradition gained press, infamy, and a loyal following after the Texas debacle in 1990 when a writing committee, under the leadership of Linda Brodkey, made diversity and difference the overt subject to be studied and written about in the required writing classes.

Although dramatically misrepresented in the press, the ensuing political correctness debate highlighted an important transition in required writing instruction: belletristic writing-about-literature and self-indulgent expressionist rhetoric was old hat; the new game was critical pedagogy. If Newkirk wanted to invigorate writing in required writing courses by having students write about subjects that were internally meaningful, the cultural studies teachers wanted students to move to the outside and write about politically and socially meaningful subjects, to resist cultural hegemony, and adopt the stance "of the interrogating individual who could recognize his/her subject position as the product of discursive conflict" (Strickland 1990, 298). But whereas Spellmeyer was championing the personal essay, cultural studies teachers appropriated various forms of argumentative discourse, adapting some of the more traditional aims of academic writing to the purpose of interrogating the dominant narratives that have naturalized in the United States the capitalist agenda.

Although it has been challenged and frequently leavened, the cultural studies model of critical thinking and writing has arguably been the dominant pedagogy of the 1990s and the first decade of the 21st century—or at least insofar as what one reads in our books and journals reflects classroom instruction (Fulkerson 2005, 60). Critical pedagogy represents in some ways an improvement over the cognitive model of critical thinking and writing: it at least works from a real exigence—although the exigence may lie more in the teachers' than in students' minds. If successful, the writing teachers who follow this cultural studies model will have improved the students' abilities to write in academic genres and also transformed them into critical thinkers who will contribute, as John Trimbur (2000) puts it, "to the unfinished business of democratic communication" (217). In this model, teachers attempt to enlist their students in the critical literacy project through reading, discussing, and writing to persuade each other to their own sense of social justice, identity formation, ideological coercion, and diversity. The critical pedagogy teachers hope their students will leave their classrooms having thought, fought, and written about serious subjects. They will see the world anew—and they will have gained this new vision as a consequence of the multiple perspectives they have been exposed to through the circulation of texts.

In many ways, I ascribe to the critical pedagogy agenda because it coincides with my world view, but I am concerned that among unpracticed teachers (and even among some of the practiced ones), it becomes another move in the social reproduction game that works by disguising stratifying strategies. This stratifying strategy is perhaps even more effective than the supposedly neutral cognitive model, because it has wormed its way into the discourse of social justice teachers by masking itself as a liberating pedagogy. Critical teachers who subscribe to this "new-vision" pedagogy may be re-inscribing social class relationships by privileging the ways of thinking, acting, and writing that are characteristic of the higher social groups and antithetical to the working-class habitus.

Other working-class academics have had similar concerns about the contradictions embedded in writing programs in which the dominant purpose is cultural critique. James Zebroski (1992) focuses in his analysis of the cultural studies program as it was too often practiced at Syracuse on the "denigrating of the student's culture and values, and most importantly, his or her ability as a member of a community to produce knowledge" (92). Although many of us pay lip service to the progressive ideals of inquiry as the privileged mode of learning and the

Freirean dialectic between student and teacher knowledge, those ideals tend to rupture when the students either refuse to learn or seem incapable of learning what we have to teach them. Along with Zebroski, I have felt that assumptions of student ignorance often seem endemic to our profession, perhaps as a time-honored method of constructing our own authority and pride in our imagined intelligence, or as Freire ([1970] 1995) pointed out, of saying our words over the words of our students (114-15), appropriating for ourselves, as dominant cultures have always done, the right to name the world.

The rhetorical situation of the teacher unfortunately impels many of us to defend our authority with claims of our knowledge. Jane Tompkins's (1990) memorable and too easily dismissed confession of her professorial insecurity in "The Pedagogy of the Distressed" excavates the logic behind our impulse to assume the pose of the knower over the unknowing. It may be additionally tempting for a professor to construct student ignorance when the student comes from a lower social class than the teacher, the student being marked by working-class dress codes, mannerisms, cultural knowledge, and language. Professors with working-class origins who have written their ways into the academic class may be even more prone than middle-class professors to wall themselves off from working-class students as a way of confirming their own membership in the privileged class and their distance from the social class of their origin.

Zebroski (1992) points out the irony of critical teachers who fall into this professorial pattern of constructing walls between themselves and their students by assuming student ignorance, in spite of the critical teachers presumably having read Freire's *Pedagogy of the Oppressed* ([1970] 1995), a central feature of which is the deconstruction of the "teacher-student contradiction" (60). True education, Freire claims, can occur *only* when teachers are in full solidarity with their students, which involves trusting them, respecting their desires, and engaging in an exchange of knowledge. After remarking on the irony of critical teachers who complain about students' unwillingness to learn, Zebroski quotes Freire warning against the ironic reproduction of difference by teachers who have imagined themselves as liberators:

> The [man or woman] who proclaims devotion to the cause of liberation yet is unable to enter into communion with the people, whom [he or she][3] continues to regard as totally ignorant, is grievously self-deceived. The convert who

3. I have inserted the dual-gender pronouns Freire used in his 1995 edition.

> approaches the people but feels alarm at each step they take, each doubt they express, and each suggestion they offer, and attempts to impose his "status," remains nostalgic toward his origins. (93)

Although Freire is referring in this passage to teachers who in non-industrialized countries have committed themselves to helping oppressed peasant cultures, Zebroski implies that the situations Freire describes can be re-read from a North American perspective in college classrooms.[4] One could translate Freire's claim here by imagining students as peasants and teachers as revolutionary educators who had renounced their social class of origin (teachers in the banking tradition). Within this framework, the "liberators" who fail to enter into communion with students by respecting their knowledge would be reverting to the system of domination they thought they had left. One might even generalize beyond these sets of homologous situations to theorize that by reinscribing the "teacher-student contradiction" (Freire [1970] 1995, 60), falsely liberated teachers are in fact reproducing the capitalist system of oppression depending on social class differences marked by those who know (the "educated" middle and upper classes) and those who don't (the "non-educated" working classes).

CENTRAL CONTRADICTIONS

Although the social strand of critical thinking travels under many names (cultural studies, critical theory, critical pedagogy, resistance theory, emancipatory pedagogy, liberatory pedagogy, critical teaching, radical teaching, transformative pedagogy, CCS, critical literacy), each with a special emphasis revealed in its name, they share common assumptions and agendas—the final purpose of which is to promote an egalitarian

4. Educators working from a Freirean pedagogy have frequently been challenged by what could be called "The North American Question," whether a theory developed to deconstruct systems of oppression in non-industrialized countries could be appropriated to the situations of college students in the United States. Freire answers this critique by insisting that people understand the relationships between generalizations and specific instances. Educators should, he says, not import his pedagogy but reinvent it within the context of the specific situation in question. The relationships between social classes, upon which his theory is built, is certainly not the same in rural Brazil and urban New York, but generalizations about social class differences and relationships can still apply to both situations marked by systems of domination (see Freire 1993, ix; Shor & Freire 1987, 24-25). Educators who attempt to copy Freire rather than reinvent him within specific situations would clearly have demonstrated their adhesion to systems of domination. The same could be said of teachers who attempt to appropriate Freire's pedagogy and then lament their students' inabilities to understand the "lessons."

social structure, concern for our environment, a global consciousness, peace over war, sharing over personal profit.

The literature promoting this project is rife with phrases like the classroom as a "site of political activity and struggle" (Berlin 2003, 122), "contact zone," resistance to "the uncritical acceptance of ideological values" (Hardin 2001, 6), "opening spaces" (Hardin 2001), language as a "terrain of ideological battle" (Berlin 1994, 478), "semiotic analysis" (Berlin 472), ideological formations, hegemony, resistance, teachers as cultural workers, borderland, border crossing, transculturation, stupidification, reproduction, demystification, cultural criticism, empowerment, emancipation, and liberation. These words and phrases are all rooted in the central activity of teaching students how to read cultural codes critically and become empowered to promote critical democracy. Berlin's (1994) description of his required writing course at Purdue has served as a model for critical teachers:

> The course is organized around an examination of the cultural codes—the social semiotics—that are working themselves out in shaping consciousness in our students and ourselves. We start with the personal experience of the students, but the emphasis is on the position of this experience within its formative context. Our main concern is the relation of current signifying practices to the structuring of subjectivities—of race, class, and gender formations, for example—in our students and ourselves. The effort is to make students aware of cultural codes, the competing discourses that are influencing their formations as the subjects of experience. Our larger purpose is to encourage students to resist and to negotiate these codes—these hegemonic discourses—in order to bring about more democratic and personally humane economic, social, and political arrangements. From our perspective, only in this way can they become genuinely competent writers and readers. (473)

I find much to admire in Berlin's description of the Purdue course as well as in his social objectives, but there are predictable problems with this model signaled by the emphasis on interpreting cultural codes, the evasion of agency in locutions like "The effort is to make . . . ," and the logical leap establishing a necessary link between good writing and egalitarian politics (a link that would have surprised Ezra Pound, T.S. Eliot, Richard Weaver, and William Buckley, Jr.). Berlin hedged with the phrase, "genuinely competent," leaving a small space for ungenuinely competent conservative writers. Locating "our perspective" in that linkage leaves another space: a place for students who don't buy the argument and who in fact may even be irritated by the teachers' attempts to impose their logic on the students who come from another social location.

In the following sections, I will focus on three central problems embedded in Berlin's mission statement: the displacement of writing instruction, vanguardism, and student resistance. Next, I will explore Freire's interpretation of false consciousness to explain how teachers committed to social justice issues slip into practices that generate student resistance. Finally, I will analyze two examples, one from an experienced and one from an inexperienced teacher, of Freire's interpretation of prescriptive methods of teaching ([1970] 1995, 107-112).

Displacement

Because many of us in rhetoric and composition share common assumptions and purposes, one might think we could approach a discussion of the balance between instruction in writing and the subject of writing in a logical fashion, but the history of the conversation is replete with charges, counter-charges, and recriminations that tilt toward ad hominem arguments, perhaps beginning with Hairston's (1990) infamous dismissal of "low-risk Marxists" (695). One might struggle to excuse Hairston for the intemperate nature of her entire comment, which I read—as did John Trimbur (1990)—as the irritation of the old guard about to leave its inheritance to a younger generation with which it had little patience. I was—and to some extent, still am—sympathetic with Hairston's outburst against Marxist-laden jargon impenetrable to any outsider, which as a graduate student still serving his sentence, I was at that time. Now that I am approaching the nether end of my career, I also understand Hairston's decision to write exactly what she thought.

John Trimbur was one of those Hairston (1990) had castigated for his "name-dropping, unreadable" prose and his "assumption that the goal of English teachers should be to indoctrinate their students in neoMarxism" (695). In truth, Hairston had a point in her critique of the style. Here is her example of Trimbur's style: "In this regard, the Habermasian representation of consensus as a counterfactual anticipation of fully realized communication . . . ," a noun phrase that must have given Joseph Williams the shivers. Like Hairston, I would be disinclined to wade through prose like this, even though I might have been interested in what Trimbur (1989) had to say about consensus and collaborative learning.

Trimbur's response (1990) to Hairston was remarkable for two reasons. First, his style became far more readable than it was in the 1989 article about which Hairston had complained. Here is an example of his more readable style: "I see," he writes, "writing and reading as powerful tools for students to gain control over their lives and to add their voices to the ongoing debate about our communal purposes" (700). Granted, the subject and rhetorical situation of the response is quite different from his article; nevertheless, in this five-page response, he takes up some complicated issues, one of which is that in spite of his demurrals, he thinks he should help his students see through the many ways in which they have been indoctrinated by the dominant culture.

More remarkable than the clarity of his writing, however, was the restraint he showed in his response. He compliments Hairston (they clearly knew and liked each other—or at least did, until Hairston wrote her 1992 article), and then proceeds to lay out in a balanced tone a careful analysis of why Hairston was annoyed and a justification for his project of "linking literacy to democratic aspirations," code for helping students see what's wrong with our culture and what they can do about setting it right. Only at the end of his response does Trimbur let fly with a zinger when he implies that Hairston is "redbaiting" (700).

Although in a more subdued tone (this time, spelling Charles Paine's name correctly),[5] Hairston (1992) fired her penultimate salvo in "Diversity, Ideology, and Teaching Writing." This article from the former Chair of CCC and a recognized leader in the field may have done more than anything else to set a confrontational tone between teachers and scholars who precipitously found themselves on different sides of the fence Maxine built. Although I side with Hairston, I wish she had been more forgiving of those who have, like Dale Bauer (1990) and Ron Strickland (1990), forcefully claimed that teachers have an obligation to challenge their students' political perspectives. Although there are manifold subtleties in the argument, the focus of Hairston's complaint and the consequent rejoinders was that teachers have let the imperatives of a cultural studies agenda override their responsibilities to help students with their writing.

Subsequent to her article, Hairston was routinely challenged and sometimes ridiculed by critical teachers. I remember a well-attended and particularly vitriolic round table session at the 1992 Rhetoric Society

5. In her Comment (1990), she spelled Paine, Payne, a miswriting that perhaps testifies to the off-the-cuff tone and style of her comment.

of America Biennial Conference with Victor Vitanza, John Trimbur, Patricia Harkin, and John Schilb as speakers. It seemed as if the speakers, with the exception of Vitanza,[6] were outdoing each other in their attempts to excoriate Hairston for her denunciation of the trend toward critical pedagogy. As a second-year PhD at the time, I was shocked. Maybe shocked is the wrong word—I was fascinated. I felt as if I had inexplicably been allowed to see something obscene.

The confrontational tone was continued in the Fall 1992 issue of *College Composition and Communication.* In the Editor's Column, Richard Gebhardt described the unprecedented volume of telephone calls and correspondences he received on either side of the fence. The tone of Trimbur's "redbaiting" remark resurfaced in one writer's complaint that Gebhardt was "publicizing McCarthyist views" (quoted in Gebhardt 1992, 295). Gebhardt also quotes Cy Knoblauch imputing to Hairston "darker motives" (296) that we are left to imagine.

In the May 1993 issue, Gebhardt included in "Counterstatement" six comments from scholars who took strong exception to Hairston's claims, a response by Hairston (in which she sidesteps the necessity of responding by saying that she and her critics are in such different worlds that no conversation is possible), and two comments by writers who claimed Hairston had got it right. The proportion of Hairston's critics and defenders presumably represented the proportion of responses Gebhardt received.

Trimbur's (1993b) response is less forgiving than his 1990 response, accusing Hairston of treating students as "dupes" (249) in the face of his more challenging project of imagining "first-year composition as rhetorical education for citizenship" (248). Strickland (1993) aligns her with neoconservatives (250, 251), charges her with confusing deconstructionism with vulgar Marxism, and encouraging a pedagogy in which students are encouraged to enshrine stereotypical images of their home cultures—Texans writing about cowboy boots, male black students about basketball, and so on (251). William Thelin (1993) accuses her of a "naïve belief that classrooms can be depoliticized" (a replay of Schilb's remarks at the 1991 Rhetoric Society of American Conference) and a "patronizing predisposition towards students [that] will forever keep

6. The speakers went over their time allotments and didn't leave Vitanza time to speak. I should also add that in retrospect, I know that these scholars, all of whom I respect, were reacting situationally. They were caught up in the moment. I suspect all of them think differently of Hairston now.

them in the role of the timid student" (252). To read her critics, one would think Hairston was in the dark ages.

The Texas debacle, which Molly Ivins immortalized as the "tempest in a teapot" (Brodkey 1994, 244), lies behind this rancorous conversation. Although Hairston (1992) mentions it only in passing, it was most likely the catalyst that compelled her to write "Diversity, Ideology, and Teaching Writing." Linda Brodkey (1994) gives the most complete narrative of the events that led to the decision by the dean of arts and sciences to "postpone" (really, cancel) the course proposed by an ad-hoc writing committee chaired by Brodkey, the writing program administrator. The proposed required course, named "Writing about Difference," essentially was based on readings about issues focusing on racism and sexism and having students interrogate the various positions and develop positions of their own—a course that many of us would currently find unexceptional, even if we were to protest against some of the assumptions underlying it. Nevertheless, as a consequence of some members of the English department going public on their grievances against the course, the controversy quickly gained national attention with pundits like George Will (Brodkey, 251) and figures like Lynn Cheney (Brodkey, 247) leading the charge against the leftist cabals in English departments determined to radicalize American youth. Although many issues were at stake in the conversation, the reigning charge was that radicals were "subordinating instruction in writing to the discussion of social issues and, potentially, to the advancement of specific political positions" (Brodkey, 248).

This is the hornet's nest Hairston stirred up when she decided to go public with her severe discontent with what is now known as the "social turn" (Trimbur 1994b) in rhetoric and composition. Although two decades have passed since this initial brouhaha, the hornets are still buzzing. Because a host of other issues lie beneath any discussion of the degree to which courses focusing on reading and writing about politically charged social issues suppress a focus on writing instruction, it is difficult to pose questions about how much time students are spending on reading, how much class time is spent on arguing about the issues, and whether the social dynamics of the teacher/student relationship gets in the way of students being able to write frankly about their beliefs. I imagine a wide river with wakes created by many boats (probably towing water skiers), a current, and a contrary wind. But underneath the water are unacknowledged demons fighting some kind of primal battle that is the real source of the waves. These demons may be linked to language issues, social-class privilege, group-think, individualism, insider-outsider dynamics, "service" courses, and as Virginia Anderson (2000) has described it in one of the more profound articles I have read on the "social turn," who has property rights on composition.

Vanguardism

The urge to correct students' naive perception of the world and word implies a vanguardist pedagogy. Although Freire ([1970] 1995) warns against vanguardism (see his discussion on "Cultural Invasion," 134-38) at length, it is the charge frequently leveled at him by a variety of compositionists determined to rupture the central contradiction within Freire's pedagogy. Richard Miller's (1998) "Arts of Complicity" is one of the more aggressive critiques of Freire as a thinly veiled Leninist.

The charges Miller levels against Freire are ones I would level against critical teachers who have taken it upon themselves to correct their students' thinking about central social issues—like diversity and difference. I think Freirean acolytes who haven't taken Freire's warnings about correcting their students' perspectives are Miller's real target, but Miller (1998) paints Freire as an unbelievably naïve teacher and theorist who doesn't see that through his liberatory posturing, he is indoctrinating his students into his own thinking. Miller's tone is sarcastic: "Everybody Get in Line: Liberation and the Obedient Response" (12) is his first subheading as he sets about to pillory Freire. Although Miller acknowledges that Freire claims to investigate with the peasants the particularities of their own realities and through that investigation open up their own critical perception of the world and word (i.e., liberatory literacy), "it is hard to believe," Miller writes, that the "critical perception of the world he seeks to impart through the problem-posing method is meant to produce anything other than a new citizenry with a shared set of values" (14)—i.e., Freire's vision of the world. Miller sarcastically refers to Freire's "ever-pliable peasantry" who "come to see that they were never drunkards, but were rather oppressed. Through Freire's pedagogy," Miller remarks, "the drunken consciousness is on its way to sobering up." Freire's way of teaching, "almost magically, produces people who know exactly what to think about injustice and how it should be redressed" (14). "Thus," Miller writes,

> Freire presents the recipients of his pedagogy as coming to their own conclusions, as learning to think for themselves. He doesn't linger over the fact that all this self-motivated thinking leads his students to think exactly what he would like them to think; he doesn't imagine that, possibly, his students are mouthing his *pieties,* silently collaborating in the production of the desired public transcript and then sneaking back home where they are free to question his lessons or force others to accept them or forget them altogether. (19, my italics)

Rather than linger over the misfiring of Miller's (1998) rhetoric, I will sketch the outline of his argument that to me reads as an unintentional self-critique, and then move to the part of his argument to which I pay more serious attention—because it fits my point of view. Miller begins the article with a narrative of himself in his younger, naïve years when he "was among those swept away by Freire's vision" (11). Within a theoretical context, "vision" normally carries a positive meaning, as in people who "see ahead," but Miller is obliquely referring to a "seeing of the other world," the world that isn't as opposed to the world that is, i.e., romantic verses pragmatic (one wonders whether Miller ever was "swept away"). Miller intends to contrast Freire's unreal pedagogy (the one that doesn't work) with his own pragmatic pedagogy (the one that does). Miller is also invoking a related religious connotation in two ways: first, Miller's critique implies a Freirean cult, attractive to young teachers attempting to escape the conflicts endemic to the profession; and second, Miller's analysis consistently arcs back to the Damascus trope: Paul, traveling along the road and going about his business of persecuting Christians until he is struck by the light from Jesus. The light momentarily blinds him, but after he comes out of his "vision," he is able to reenter the real world and "see." After a period of counseling and baptism, he then retakes to the road to preach the new religion—Freireanism, or whatever cult is au courant.

Following the genre of the deprogrammed acolyte, Miller describes his disillusionment with Freirean pedagogy, linking himself to other early Freireans, who wondered why Freire's "vision" didn't work in North America (Ellsworth 1989; Weiler 1994). Miller's attempts to en*lighten* his students leads to either resistance or ventriloquated garbage. Miller's pedagogical failure leads him to question the new religion (when he might have questioned his reading of it). The primary hole in Freire's bucket, Miller says, is the contradiction between the claim to teach students how to think for themselves and the effect of teaching them how to think like the teacher. This vanguardism, Miller points out, is what makes students resist either overtly by refusal or covertly by pretending to go along with the game.

Miller insightfully analyzes why he and other teachers have been drawn to this flawed pedagogy. In essence, he proposes two reasons: First, a liberatory pedagogy offers teachers an escape clause from their complicity in the social reproduction function of education. Pretending to be deconstructing social structures through left-wing pedagogy is an effective way of accepting the privileges that go along with being a professor while attacking the structures that have afforded them that privilege. And second, rather than be toadies, they get to picture themselves as "teacher-hero[s] who [get] down to the business of liberating right, left, and center." Miller claims that a "Hollywood drama" of "conversion and redemption" is what drives these liberatory pedagogues (26). The practitioners have been raised, S(P)aul-like, to swallow whole the capitalist mythology of competition, the unitary individual, and meritocracy. They have gone to school, done a little reading, and come across gurus like Berlin, Freire, Gramsci, or Foucault. The veil having been lifted from their eyes, they set out to convert the blind perhaps as a way of confirming their own current illusion of sight.

But while criticizing Freire and Freireans as conversion pedagogues, Miller is playing the same game, which might be called a sub-genre of academic discourse—i.e., explaining how everyone else is wrong and you're right. After exploring the contradiction that implodes Freire's pedagogy, Miller offers his new way of teaching—aided by James Scott's trope of the public and hidden transcript (which to an ex-high school teacher doesn't seem like a remarkable insight). Miller calls his new method a "pragmatic" pedagogy as opposed to Freire's *vision*ary pedagogy. As Freire has done with his "ever-pliable peasantry," Miller trots out examples of his students who have been en*light*ened by his "more modest" goal (27). Rather than liberate, he is merely trying to give them "discursive versatility," which in essence means learning how to read different kinds of rhetorical situations and respond appropriately—or as Alan France (1993) postmodernly puts it in another context and from an entirely different angle, "extending the range of discursive entry-points assigned to students" (594).

Although Freire is obviously not as unreflexive as Miller sets him up to be, I agree with Miller's critique on several counts. First, Freire's theory is built on a contradiction of free thinking that leads to a predestined conclusion. Freire's central assumptions is that when people learn to think critically, they will perceive that to be human is to be able to participate in the transformation of the world and not to block anyone else's ability to participate in that transformation ([1970] 1995, 38-39). Freire struggles to present the tie between thinking critically and this perception as a deduction, but to me (and to Miller) the assumption is a priori. Although I'm no philosopher, I suspect one could just as carefully reason one's way to nihilism.

But as Zebroski (1992), George and Shoos (1992), Frey (1998), Mack (2006), and other serious Freirean scholars realize, Freire does not come to the peasants to export his conclusions.[7] Rather he has faith that thinking critically will lead people to see that "the important thing is the continuing transformation of reality, in behalf of the continuing humanization of [people]" ([1970] 1995, 73). The flip side of Freire's logic defines the oppressor as one who restricts someone else's freedom to work on behalf of that transformation. Through dehumanizing others (limiting their abilities to be fully human), the oppressors "themselves also become dehumanized" (38). A priori or not, I tend to agree with Freire: *if* our purpose is to promote living with others and with the world, then we have to work from that base of participation and non-oppressive praxis, which is far better than working from a nihilistic base in which all is permitted.

Secondly, I tentatively agree with Miller's speculation that writing teachers are drawn to liberatory-like pedagogies as a way of pretending they are undermining the system from which they are gaining advantage. This kind of liberatory teacher may well fall into the vanguardist trap as a consequence of her desire to obscure her own complicity. But the motivations drawing critical teachers to Freire are surely more complex than Miller's reading of them implies. Their motivations may differ as a consequence of their race, gender, social class origins, and

7. In "Critical Pedagogy: Dreaming of Democracy," Ann George (2001) gives a thoughtful survey of the various interpretations of Freire's pedagogy. She makes clear in her survey that one of the central conflicts has been the degree to which emancipatory educators are simply replaying banking style education by depositing their knowledge into the students' accounts. Responding to this charge, George and Shoos (1992) quote the following from Freire: "The sectarian, whether rightest or leftist, sets himself [sic] up as the proprietor of history, as its sole creator, and the one entitled to set the place of its movement. . . . They are similar in imposing their own convictions on the people, whom they thereby reduce to mere masses" (201).

trajectories. The motivations driving a white male professor with middle-class social origins and a rapidly rising professional trajectory might be quite different from those of a black working-class woman who didn't start undergraduate school until her early 30s, who was a single mother with two children, and who didn't finish her PhD until her late 40s, who is teaching in a community college, and who has not made it big on the rhetoric and composition scene.

Finally, although I would put it differently, I agree with Miller's purpose of helping students gain "discursive versatility." My difference with Miller is that I have come to my purpose through Freire while Miller comes to his pragmatic pedagogy against him. Freire consistently warns against the vanguardist in sheep's clothing that Miller takes him to be. Freire insists on a careful investigation of the students' worlds, on respecting those worlds, and at arriving at a literacy agenda with the student. Dialogism is fundamental to this approach. Good teachers *always* work and learn with their students. This means in the required writing class that teachers should investigate their students' literacy skills and goals, honor them, and work with them to help them improve their skills and reach their goals, even though their goals may be quite different from the ones the teachers had in mind. To me, this seems likes good teaching.

Student Resistance

Even among practiced teachers like Berlin (1991a) and Bauer (1990), stories about student resistance to critical pedagogy are commonplace. After an extensive review of the literature, Joe Hardin (2001) notes that "those who are committed to the projects of critical literacy often report as many failures as successes in their attempts to adapt resistance theory and critical pedagogy to their classroom practices" (98). In an endnote, he adds that these reports of failure have to be interpreted within a reflexive ethos because reflexivity is built into the critical teacher's praxis. To support Hardin's claim, we should note a book edited by William Thelin and John Paul Tassoni (2000), *Blundering for a Change: Errors & Expectations in Critical Pedagogy*, a collection of stories about blunders in critical literacy classrooms. To my knowledge, *Blundering for a Change* is the first book that posed a willingness to be frank about our failures. None of us like to publicize what we did wrong; the culture of publication in fact argues against such exposure. The ethos of self-presentation requires accenting in sly ways what we do well, even when narrating, as Miller does, our pre-enlightened mistakes. Unsurprisingly, Thelin and Tassoni's book has been an undeserved bomb on the market, ranking 2,159,864 in the Amazon ranking, to date.

Although there may be something to Hardin's claim about critical teachers being more willing than non-critical teachers to be self-critical, this claim implies that "non-critical" teachers like Maxine Hairston, David Schwalm, Paul Anderson, Joseph Harris, Jeff Smith, Joseph Williams, or Gregory Colomb don't reflect on their own practices. We should also note Elizabeth Ellsworth's (1989) charge that after her review of the literature, she concluded that one of the central problems of critical pedagogy is precisely its *lack* of reflexivity. According to Ellsworth, the critical literacy narrative generally imagines naïve students who have been brainwashed by democracy's master narrative, aided by uncritical teachers who perpetuate the old stories about individuality and the free market of ideas. She cites Peter McLaren:

> [T]hese students do not recognize their own self-representation and suppression by the dominant society, and in our vitiated learning environments they are not provided with the requisite theoretical constructs to help them understand why they feel as badly as they do. Because teachers lack a critical pedagogy, these students are not provided with the ability to think critically, a skill that would enable them to better understand why their lives have been reduced to feelings of meaningless, randomness, and alienation. (311)

Into the arena enters, as Miller put it, the teacher-hero, who because of her exposure to Marx, Althusser, Foucault, Hall, Giroux, et al., has a truth that is her duty to impart (as well as whatever else she was supposed to be teaching) in the hope of encouraging her students to be foot soldiers in the transformation of a sexist, racist, and classist culture.

But she meets resistance. The reason is not her misteaching but her students' submersion in hegemonic discourse: Hardin (2001) puts it like this: "A problem for critical pedagogy, as many radical scholars and teachers have noted, is the *fact* [my italics] that a great many students already may have been fully acculturated into dominant culture by the time they arrive in the composition class" (49). He cites Berlin's interpretation of the "'stiff unwillingness' on the part of students to question and problematize the cultural values presented by literary or other texts" (30).

As a consequence of my reading of resistance theory narratives, my practice, and my close relationships with many critical teachers, I tend to agree with Hardin that critical teachers are open to self-criticism, but not to any significant degree more than non-critical teachers, particularly if I include in this category the non-liberatory teachers I mentioned above. Like all other teachers, critical teachers resist looking too closely in the mirror (Ellsworth's point). It's always a lot easier to blame the "out-there" than fix the "in-here." And it's always a lot easier to blame the students for one's own misteaching.

Although I shouldn't be, given the master-narrative of dominant-class knowledge over dominated-class ignorance, I am continually surprised by the durability of the student-ignorance story, in critical and non-critical (if there is such a thing) teachers alike. No matter how often teachers like Dewey ([1938]1963), Freire ([1970] 1995), Moffett (1968), Ellsworth (1989), Zebroski (1992), Mack (2006), and I (2003) continue to protest this logic, it keeps popping up like the seven-headed hydra. I think it's possible that the stories of student resistance, of which I have experienced my share, have more to do with the teacher/student narrative than it does with any particular pedagogy. Our challenge, then, has less to do with critical or non-critical teaching than with learning how to respect our students' lives.

FALSE CONSCIOUSNESS

In spite of Miller's portrait of Freire, Freire has consistently insisted on the necessity for fixing the "in-here" before trying to correct the "out-there." His analysis of the problems with teachers' motivations spotlights why students resist critical pedagogy. Freire ([1970] 1995) reminds us that middle-class reformers "who espouse the cause of liberation are themselves surrounded and influenced by the climate which generates the banking concept, and often do not perceive its true significance or its dehumanizing power. Paradoxically, then, they utilize this same instrument of alienation in what they consider an effort to liberate" (60).

By "climate" Freire is referring to the social structure of domination in which the dominant classes have been conditioned to think of the lower social classes as lacking education and initiative. Having grown up within this structure of domination, the middle-class reformer tends to believe that to "raise" themselves up, members of the working-classes need to be "taught" literacy and other social/moral skills—the ones that Lynn Bloom (1996) speaks of when she describes Freshman English as a middle-class enterprise.

In addition to inheriting a hierarchized notion of literacy and values, middle-class reformers have spent most of their formative years in school, which reinforces through the teacher/student relationship the middle-class/working-class contradiction, within which is embedded Freire's concept of false charity. Freire opposes false to true charity, the latter requiring one to be in full solidarity with the oppressed. To be truly solidary with the oppressed, a reformer must escape the structure of domination, which both the social order of domination and educational institutions reproduce. The escape isn't easy. I am personally still tied with ropes, if not in chains; being tied up, however, does not block me from recognizing the possibilities.

Lest academics with working-class origins applaud themselves for not having been born into the oppressor classes and consequently for being free from the influences of the social structure of domination, Freire ([1970] 1995) theorizes a correlate condition for intellectuals who come from the oppressed classes but who do not break free from the teacher/student, dominator/dominated contradiction. In brief, people from the oppressed social classes internalize the contradiction that is the foundation of the structure of domination. Freire's interprets this internalization phenomenologically, as if the oppressors enter like specters into the mind of the oppressed. The oppressed see themselves through the internalized specter's eyes. They try to re-establish their lost humanity by becoming the specter and putting someone else as a surrogate in their places.

Through a process of adhesion, working-class academics may work through the system to become a working-class version of the oppressor. They gather the necessary degrees to become "teachers." Although they may have imagined that they would be teachers-with-a-difference, they will be inclined to internalize the habitus of the teacher, to assume the teacher authority and position that puts the teacher on one side and the student on the other of the academic divide, an externalization of their own ambiguity. "The very structure of their existence," Freire ([1970] 1995) explains, "has been conditioned by the contradictions of the concrete, existential situation by which they were shaped" (27). A fundamental element of the existential situations academics inhabit is prescription. "Every prescription," Freire says, "represents the imposition of one individual's choice upon another, transforming the consciousness of the person prescribed into one that conforms with the prescriber's consciousness." (One wonders whether Miller thought Freire didn't *understand* what he was writing?)

On a larger scale, the dominant elite (see especially Bourdieu's analysis 1984, 270) make the dominated fraction of the social contradiction see both themselves and the world through the eyes of the dominating fraction; this is a phenomenological explanation of hegemony, into which the critical teachers who have failed to resolve the teacher/student contradiction fall. They have jumped through the hoops legitimating their knowledge, their self-imagined clear-eyed vision. Institutionally legitimated, they now come to the classroom intent upon liberating their students who, to their minds, have been indoctrinated by capitalist hegemony.

It should be no surprise that some students resist prescriptory teachers, because the central effect of the social structure of domination is alienation. The students are treated as objects who do not know, a few of whom become subjects only when they absorb the knowledge of the teacher, at which point, they pass over into teacher and busily reproduce the structure of domination that produced them. The alienation is the consequence of being treated as outsider to the in-group of knowers. The irony is that this kind of prescriptory pedagogy is masked as teaching students how to think critically. You can always tell teachers who have been absorbed by "structuring structures" of the educational institution (Bourdieu 1991, 168-170)—you will hear them remark on students' inabilities to think critically, implying that only a critical thinker can perceive non-critical thinking.

PRESCRIBING CRITICAL THINKING

In 1990, Ronald Strickland published an article predicated on the knowledge divide. The teacher knows because he or she has been doing all this sophisticated reading, writing, and critical thinking about subjects like—in Strickland's case—*Paradise Lost* and its social function as a misogynist text. Strickland's interpretation of a student's (or anyone's) ignorance is interesting. The ignorance isn't, in Strickland's view, entirely the consequence of the student's lack of a sophisticated language (although this lack plays its part insofar as the student does not have the vocabulary and syntax to name counter-hegemonic practices); rather, Strickland ties the student's lack to Lacan's theory of ignorance as repression, a staving off of that-which-one-does-not-want-to-know. Strickland links the Lacanian notion of ignorance to the Marxist/Althusserian interpretation of ideological interpellation: i.e., the category of undesired knowledge is a consequence of the socializing mythologies and practices in which students are submerged. According to Strickland, the dominant culture determines desired and undesired knowledge to serve the "corporate-sector demand . . . that shapes students to fit the needs of a capitalist and patriarchal society" (291). Strickland imagines that students have been indoctrinated by the late capitalist system; consequently, his "confrontational pedagogy" is designed to shock them out of their resistance to counter-hegemonic knowledge (see also McGee 1987).

Because the teacher is knower and students are non-knowers, they are inescapably, according to Strickland (1990), in conflict. Strickland's rationale for entering into an openly confrontational relationship with students predicts the conversation that followed Mary Louise Pratt's 1991 article, "Arts of the Contact Zone," Gerald Graff's *Beyond the Culture Wars* (1992), and Patricia Bizzell's (1994) use of "contact zone" as a metaphor for reconfiguring English studies. Pratt theorized the "contact zone" as a place where representatives from different social groups with asymmetrical power relationships meet, each coming with their different habitus, and negotiate, in an approximately common language, their differences. Strickland is refreshingly forthright in his objectification of the asymmetrical power relationship in the contact zone of the classroom. Although he interprets the asymmetry primarily as a matter of knowledge versus ignorance, he mentions almost as an afterthought the asymmetry contingent upon the relationship between grade-giver and grade-receiver, glossing over the fact that the teacher's power is also a consequence of his or her having been born or assimilated into the correct social class and/or having accumulated the appropriate symbolic capital to legitimate his or her knowledge. Strickland acknowledges that by positioning teachers as inevitably in conflict with students, he risks an authoritarian pedagogy, but he writes that his insistent challenging of students' positions creates the classroom as "a discursive site in which knowledge can be produced" (292) rather than mindlessly replicated.

Strickland's (1990) strategy is a replay of the Iowa writer's workshop model: he creates a situation in which students familiarize themselves with contending ideas on a certain issue and then write position papers on it. The papers are reproduced in class sets. The class members will read a paper and respond to it in what is essentially a whole-class attack, which Strickland as teacher joins. Strickland is very clear about avoiding the more gentle models of "assimilation and apprenticeship" (295). The gloves are off—this is an adult game in which students are expected to be informed, take positions, and develop coherent, substantiated arguments to support their claims. Strickland hopes that by confronting students and by getting them to confront each other, he will force them out of their active resistance to new ways of seeing what the system has taught them not to see.

In many ways, I admire and in a moderated fashion have replicated Strickland's (1990) process. I agree with his primary assertions. Students do enter our classrooms fresh out of late-capitalist hegemonic school. Many of them believe in the myths of individuality, democracy, meritocracy, and god. They want to have teachers spoon-feed them knowledge so they can get their As and assume middle-class lives, such as the ones Strickland and I lead. It's also tempting to coddle students, knowing full well that most of what they say is full of middle-class, adolescent tropes.

Although we in English studies are more inclined to criticize the philistines in business and engineering school, Strickland (1990) indicts as well professors and students who think they are producing new knowledge while swallowing all the old stories about culture, literature, and the muse, imagining themselves as something special because either they are writers or they are writers writing about writers (or postmodern writers writing about writers writing about writers). Strickland wants to explode the myths by turning issues inside out for his students, by interrogating the relationship between student and teacher, and even by disrupting the reproductive function of grades.

As might be expected, Strickland (1990) meets student resistance. Unable to escape entirely his liberal arts bias, he points in particular to "students who simply seek to fulfill a humanities requirement as painlessly as possible and who are eager to continue their preparation for business and professional careers unencumbered by an encounter with a teacher committed to feminist consciousness-raising. These students make it difficult for me to pursue my political/intellectual agenda" (297). The conflict game turns serious: Strickland's agenda was to teach students how to resist late capitalist, white male hegemony, but the students resisted him, turning resistance theory inside out.

The gun backfired for obvious reasons. Strickland (1990) pictures students as ignorant because of their uncritical resistance to any theories questioning capitalist mythologies, ignoring that in each student lies a different mode of resistance to the stories his or her culture, social groups, parents, and peers narrate. Many students don't buy uncritically the myth of meritocracy any more than they buy the myth about the teacher who knows and the student who doesn't. Certainly we all remember our early school days and the not-so-gradual awakening that some teachers were screwballs. And the worst screwballs—at least as I remember it—were the teachers who with a certain smugness thought their students were stupid. Teachers who come into the classroom with cognitive chips on their shoulder will obviously meet student resistance. That chip comes in many disguises, but students see through them, whether the chip is cloaked in knowledge of literature, grammar, writing, epistemology, philosophy, or the social construction of identity. It's still the old game of I know and you don't. I've got the grade and you need it. I'm in control here, and you're not.

Freire ([1970] 1995) claims that until teachers have been able to resolve the socially constructed student/teacher contradiction (the one that is at the heart of Strickland's confrontational pedagogy), there will be no true education. Rather than work from a premise of student ignorance, Freire insisted on working like Dewey ([1938] 1963) and Moffett (1968) from the base of what the student knows. A corollary of dialogic education is that students know things teachers don't. The exchange and merging of knowledge is what makes education work. Any pedagogy that works from the prescriptory model is bound for trouble; prescriptory teachers will inevitably get angry at students who make it impossible for the teachers-as-knowers to pursue their agendas. These teachers are also headed for burn-out. But teachers who work from the Freirean base of a serious investigation of and respect for the students' worlds and their knowledge will never want to retire.

I have aimed at the vanguardist ethos that dominates Strickland's (1990) essay, although this is certainly not the ethos Strickland intended. Like all well-meaning teachers convinced of the value of the knowledge they have to impart, Strickland clearly would prefer the image of Socrates to Lenin. He wants to teach students how to think critically about cultural mythologies and how to argue their positions—the fundamental objectives of the social strand of critical thinking pedagogies. He imagines himself as open to critique and change, but his foundation of a teacher-knowledge/student-ignorance opposition locks him into a Leninist position, the intellectual who has Truth to give to the masses.

Strickland (1990) was one of the first to stake out the territory of a confrontational pedagogy. Strickland is a sophisticated theorist who advances claims that I believe we must take seriously, albeit ones that lead to dysfunctional moments in the classroom as a consequence of the knowledge divide. In spite of the references to student resistance in his classroom, I sense a lot of good teaching going on in Strickland's classroom, but I want to show how this confrontational stance can lead to disastrous classroom confrontations.

In 1993, Jeff Smith cited an extraordinary writing task constructed by a graduate student teaching assistant. Because it dramatizes the conflict of the teacher/knower and student/not-knower, I will cite it here in full:

> O.K. So here we are at the end of the quarter and I am totally exhausted from trying to inspire critical thinking and creating in you the desire to question yourselves and the world. Honestly, I think I've failed. . . . I've tried to shake up your own "ideologies" and points of view. My attempt was to have you really look at yourselves and the world(s) you live in, not to just accept them and reject new and different ones. But it seems to me that it is impossible to fully question or criticize anything if you are happy, content, and comfortable. None of you have ever starved, or been shot or bashed by anti-gay people; few of you have ever been called "nigger," "spic," "Jew," "chink," or "faggot"; and all of you have never had a reason to question anything seriously because from your safe positions in middle-class America, you cannot see the problems, lies, and sufferings of millions (which you are connected to and responsible for). Indeed, to not question or be aware of all this is to keep it happening, and to support it! Those of you who do see these things either blame the victims or just don't really care. So, for your final paper, I want you to become a victim of ideology; pretend (if possible) that you do not have the safety net of your way of thinking which is so in line with dominant thought; pretend, in short, that this society was actually hurting you by way of media myths, historical myths, and the various "Truths" offered in the Bible, the Koran, *Beverly Hills 90210*, government, commercials, history books, and language. Using the various texts read for this class, try to figure out how or what you would think about the world if it all seemed aimed against you as a free-thinking person and that thinking would save your life.
>
> Rewrites due on Nov. 27; final drafts due on December 6. (731-32)

Although Smith interprets this remarkable outpouring of teacherly discontent as a marker of expressivist pedagogy, I read it as a barebones signifier of what lies behind the social-epistemic language/knowledge dialectic. The student has been interpellated (Berlin 1988, 479) by democratic, capitalist, individualistic mythologies. The graduate student teaching assistant with the aid of various readings (one suspects *Ways of Reading* [Bartholomae and Petrosky 1993]) has been struggling "to shake up [their] 'ideologies' and points of view," just as Strickland

(1990) had been trying "to make it as difficult as possible for them to get what they expect from the course, to confront and contest students in ways that will challenge them to recognize and rethink their assumptions" (297). Strickland was far more sophisticated than the graduate student and consequently seems to have alienated only portions of his classes, but the graduate student has managed to turn his or her class entirely against him or her. The students have resented the teacher's assumption of their ignorance. The consequent classroom dramas are painfully clear. We have students who haven't learned very much about writing and who have reified their resistance to resistance pedagogy. The teacher in a fit places them in a painfully impossible rhetorical situation that will predictably result in essays that confirm the teacher's interpretation of student ignorance. A lot of mis-teaching is going on here in the name of critical teaching.

It would be reassuring to think that this kind of "pedagogical terrorism," as Giroux (1993, 50) names it, is only the consequence of an overly excited novitiate having recently discovered his or her situatedness, but I frequently hear the same complaint about student ignorance in faculty lounges, on listservs, in conferences, and in journals (see Stanley Fish's [2002] characterization of student knowledge). Even Berlin, an exceptionally skilled theorist and teacher, admitted that his agenda of desocializing his students "is never pursued without some discomfort" to his students at Purdue (2003, 112-113). His passive locution was disingenuous, but it still contained traces of the devalorization of student voices that found its most naked expression in the teaching assistant's outpouring of pedagogical grief. More tangible was Patricia Harkin's complaint at the 1992 RSA session with John Trimbur, John Schilb, and Victor Vitanza over her students' resistance to recognizing how capitalism, through the agency of language, had situated them. Harkin interpreted this resistance as evidence that first-year students had not yet reached the maturity necessary to participate in a writing course requiring serious critical thinking.

I assume that Strickland, Harkin, and the anguished GTA have moved beyond their interpretations of student ignorance, but my point is that the social-epistemic assertion of the dialectic between (written) language and knowledge coupled with a revolutionary fervor leads toward this kind of teaching, particularly among new teachers who have recently been enlightened. One of the most common assumptions of my new GTAs at Louisiana State University is that their job is to teach students how to think critically by having them read articles and then argue in writing about subjects like diversity, gender and sexual orientations, political action, poverty, and ecology. Few support an instrumentalist approach of teaching students writing strategies that will help them survive undergraduate school and succeed in their after-school professional lives. To assume a service orientation is like asking them to teach technical writing instead of the traditional second semester course in argumentation—an opposition that has its roots in the conflicting habitus of the working-classes and the elite. At the risk of adopting the same double-enlightenment position I have critiqued in Miller, I will note that I remember clearly when I thought teaching technical writing was beneath me. I wanted to work with ideas. Needless to say, I was social climbing.

7

THE TEACHERS

In chapters five and six, I have analyzed the ways in which pedagogies based on the two strands of critical thinking can lead to counterproductive teaching in required writing classes—if one assumes that the primary purpose of these classes is to help students improve their writing abilities within a family of genres. I take this purpose one step further in the writing program I direct. I ask teachers to focus on writing strategies that will help students cope with writing tasks they are likely to meet in other undergraduate courses—strategies we have determined by analyzing specific writing assignments teachers in a variety of disciplines have sent us.

Our focus on preparing students for other undergraduate writing tasks places our program squarely within the instrumentalist camp. Largely as a consequence of Spellmeyer's *Common Ground* (1993) in which he maligned instrumentalism as an "accommodation to the existing social order" (8), instrumentalism has displaced current-traditionalism as the dirty word of composition. Any pedagogy that *merely* teaches students how to fit into existing social structures is necessarily in conflict with a transformative pedagogy whose practitioners are committed to creating a more just and equitable society than the one in which we now live.

In one sense, the opposition between teachers who embrace and those who reject instrumentalism is quite simple. Some teachers are willing to interpret writing as an "instrument" students can use for their own purposes. Others, as one of my favorite graduate students told me, have to do more than "just teach writing." Amy Lee (2000), a committed transformationalist, also writes of her need to "promote something more through my pedagogy, something other than better writers" (33). That something is her interpretation of a better world.

As if it were a hinge, instrumentalism seems to swing accommodationists against social transformationalists, but the conflicts embedded in the word are not quite that simple. "Instrumentalism" is a supersaturated word, loaded with conflicting intents, accents, prejudices, and ideologies, all pushing against the word's membrane. The tensions within the word sometimes account for ruptures even among teachers who would normally be housed in the same camp.

I imagine, for example, the tension in "instrumentalism" between materialism and its unspoken antinomy, aestheticism, the former aligned with the working classes, the latter with the upper classes (Bourdieu 1984, 34, 48, 53-54). The working-class habitus plays close to material reality and the value of labor in real terms of food, rent, and clothing. They accept instrumentalism, the notion that one thing is used as a "tool," to gain other ends. Their lives are filled with tools that help their labor. They are also aware of the degree to which they are used as tools to serve the pleasure of the ruling classes. Instrumentalism is aligned with function—the value of an instrument (or person) is determined by the work it performs.

Aestheticism, marking the upper classes, is distinguished by its distance from labor. Members of the upper classes are "above it all," in the realm of intellect where thinking is removed from material necessity (Bourdieu 1984, 53-56). In the academic world, the opposition between aestheticism and materialism would be between theoretical and applied [name your field]. Within English studies, we can see this opposition played out between those who teach students how to interpret literature and those who teach students how to write.

The theoretical impulse is associated with the upper classes and the pragmatic with the working classes. Thus, one can interpret within a field, different gravitations linked to different social class orientations: teachers are working-class; theoreticians are upper. When they are really good in their field, theoreticians rarely have to teach. My research on the relationship between social class and institutional category shows, unsurprisingly, that working-class academics are disproportionately located in community colleges (34% working class; 66% middle class) and middle-class academics in doctoral extensive universities (20% working class; 80% middle-class). It is also no surprise that academics who publish frequently are rewarded far more than those who "just teach writing." This oppositional reward system is an expression of our culture that rewards non-labor disproportionately more than labor. This asymmetrical distribution of rewards is an articulation of Bourdieu's thesis that the further you get away from labor, the more you make.

If we accept a relationship between materialism/aestheticism and lower/upper classes, we can interpret anti-instrumentalist as an ironic expression of social climbing within the intellectual fraction of the middle classes—i.e., middle-class academics. The critique of instrumentalism is aimed at the economic fraction's use of others as instruments for one's personal gain (Seitz 2004, 55), or as Bourdieu (1984) puts it, appropriating the symbolic over the real. For members of the middle-classes, who are forever aspiring to be more than they are, people do not have existential value; rather, they have value as tools to be used for one's advancement (Bourdieu 252-253). Indeed, the self is only an image that one projects with one's eye on the audience as to how the image is received. Everything, that is, is only an instrument for something else.

But by flying their anti-instrumentalist flag, members of the intellectual fraction of the middle-classes are enhancing their own social status within the intellectual fraction, which exists in a dialectical relationship with the economic fraction; as you knock down the values of those whom you interpret as being in opposition to your social group, you rise in your own. This expression of anti-instrumentalism, however, contradicts the values of middle-class intellectuals who have embraced a social justice agenda because their stance places them in direct opposition to those whom they propose to rescue. By derogating instrumentalism, the middle-class intellectual is assaulting a working-class student's home Discourse. The working-class student, perceiving the attack on the values with which he or she has grown up, has to face the choice of anomie or resistance (Berger and Luckman 1967, 155-156) leading to the internalization of social class conflict, which is far from a safe space within which to write.

Finally, to complete the complication of the wars within the word, each social class fraction embraces instrumentalism within one frame but rejects it in another. The working classes, for instance, reject the seeming-over-being that marks the economic fraction of the middle class. Neither the working nor the upper classes put on fronts. What you see is what you get—the working classes, "who do not have this concern with their being-for-others, and . . . the privileged classes, who, being sure of what they are, do not care what they seem" (Bourdieu 1984, 253). The working classes resoundingly reject the instrumentalism involved in using others for social gain, but they adhere to the "tool" mechanism embedded in the educational project—i.e., getting a degree in order to get a better job. One might speculate that working-class instrumentalism embraces the use of de-personalized objects but rejects the use of others or false images of themselves to climb up the socioeconomic ladder. The economic fraction of the middle classes embraces the full use of instrumentalism, using anything as a tool to gain something else. And the intellectual fraction of the middle class embraces the use of anti-instrumentalism as a tool for their gain within the subculture defined by their anti-instrumentalist stance, i.e., those marked by what Bourdieu calls the aesthetic disposition (Bourdieu 28).

"Instrumentalism" is thus more complicated than its critics have suggested. A naïve rejection of instrumentalism might in fact be one of the reasons critical pedagogy goes north when it thinks it is heading west, an ironic expression of Bourdieu's theory of social groups struggling for control of the symbol systems structuring social relationships (see Swartz's analysis 1997, 117-129). Intellectuals, perhaps imagining themselves as representatives of the dominated classes, are aiming for dominance over the economic fraction of the middle classes, but they hit the working classes.

Critical teachers' attitude toward their working-class students' instrumentalist goals is key to the well-documented student resistance to the social strand of critical thinking, particularly among the working-class students, who are less likely than the middle-class students to interpret education, like argument, as game.

CRITICAL THINKING IN ENGLISH 102

In *Collision Course*, Russel Durst (1999) gives an extended example of student resistance to the social strand of critical thinking, resulting in misdirected teaching in spite of a talented and committed teacher and students who wanted to improve their writing. Durst's ethnography shows us how an optimal teaching situation can be undermined by the required readings and the perceived political bias of the teacher. Joshua, an upper-middle-class engineering student, complained that the readings and teacher were trying to "cram these liberal ideas down our throats" (157), a familiar complaint Durst claims was shared by the other students as well. Durst, with his own bias, labels Joshua a conservative and thus a student whom left-oriented teachers might dismiss as interpellated by the dominant ideology, but Durst's description of Joshua (23-25) paints a picture of a student many teachers would enjoy having in their courses—a strong writer, a reader, an intelligent, middle-class student who knows where he wants to go and the instrumental function of education in helping him get there.

The class offended not only Joshua but also Louise, a conservative, non-traditional, working-class student with very specific instrumentalist goals. "I hated 102 from the start," Louise said, "And then all those things we talked about from the textbook, the political stuff. I thought there was no point to it at all. It was irrelevant to what I needed to be in college for, so I lost patience right away" (120).

By "right away," Louise may have meant while she listened to the teacher, Sherry Stanforth, a third-year graduate student, read aloud from the syllabus her objectives for the course. Even Stanforth's language would seem strained to a working-class student used to straightforward syntax and diction: "to explore, with a challenging eye, the cultural forces which have shaped you . . . analyze, argue and ponder perspectives regarding the individual, the family, progress and opportunity, race, gender, education, media and democracy." Stanforth's style here is studied—"with a challenging eye" and "ponder perspectives"—creating a tone of distance between the writer and her text, as if the writer had thought too long about the words before writing them down. I can feel Louise's misgiving as she listens to this artificial language through which she learns this is a course about values—and, perhaps, writing.

After this surprising and rather stiff introduction to the class, the students were introduced to their text, *Rereading America.* As Durst (1999) notes, both the title and the introduction immediately contributed to a negative attitude in the class. The students didn't think they needed to re-read America. Like Louise, they may have thought they needed instead to work on their writing.

Their first assignment was to read the introduction, "Thinking Critically, Challenging Cultural Myths." I have read this introduction in two ways. For one, I read it from the position of an academic who has learned to agree with almost everything the editors say. The editors are explaining in generally accessible language some of the fundamental concepts behind social epistemicism. The editors don't make radical claims; they try to be fair-minded to avoid alienating student readers who haven't thought about the ways in which notions of family, equality, wealth, and democracy are myths governing behavior and identity, largely for the benefit of dominant social groups.

I have also tried to read this text as if I were Louise, a re-reading that takes me back to my identity several decades ago. I can't quite read this way, but I can get close enough to see words I don't like—"ponder" being one of them. I imagine Louise being puzzled by words like "infallible," "ethnocentric," "dichotomies," and "acculturation." She would perhaps be irritated by the condescending tone in the first paragraph in which the writers describe what a beginning college student might feel with the "pleasures of independence," the "increased temptation, and a whole new set of peer pressures." A mother of three, Louise certainly would reject the assumption that as a new student, she might be living in a dorm with "people whose backgrounds make them seem foreign and unapproachable."

If I had been Louise, I would have been offended by the writer's implication that as a student, Louise had been "blind[ed]" by cultural myths, for example the myth of family as part of a web of myths that had "enmeshed" her in a naïve way of seeing things—and that the purpose of the book was to help her re-see, to see critically, to question everything she believed in—like family, democracy, and economic mobility.

As evidence of students' need to see in other ways, the editors offer an example that I found both misleading and patronizing. They compare a professor's working-class students' interpretations of Theodore Reothke's "My Papa's Waltz" to his supposedly more sophisticated interpretation. The poem (qtd in Colombo, Cullen, and Lisle 1992, 4) reads:

> The whiskey on your breath
> Could make a small boy dizzy;
> But I hung on like death:
> Such waltzing was not easy.
>
> We romped until the pans
> Slid from the kitchen shelf;
> My mother's countenance
> Could not unfrown itself.
>
> The hand that held my wrist
> Was battered on one knuckle;
> At every step you missed
> My right ear scraped a buckle.
>
> You beat time on my head
> With a palm caked hard by dirt,

> Then waltzed me off to bed
> Still clinging to your shirt.

The editors remark that a professor at a Los Angeles City College (i.e., one of the editors) interpreted this poem "as a clear expression of a child's love for his blue-collar father." The blue-collar students, however, read the poem in a darker vein, interpreting "My Papa" as "an abusive father and a heavy drinker" (4-5). The editors claim that the blue-collar students missed the "complexity" of the poem, "the mixture of fear, love, and boisterous fun" (5). I am reading this poem framed by the memory of my father, a borderline alcoholic given to threats of violence. I am also imagining Louise's frame of her childhood home that she described as dominated by frequent "yelling, and hitting, and screaming" (Durst 1999, 140). I think of Dorothy Allison's (1993) fictionalized autobiography, *Bastard out of Carolina*, and other personal narratives of working-class life in which the father was the provider, working in poorly paid jobs in which he was treated as only an object of labor, and finding his solace for broken dreams in the community of the neighborhood bar. Read through narratives like these, myth or not, I don't see the child's clear expression of love; I hear instead a dark, potential danger in that poem, and I wonder what else the child is clinging to. If Louise read the poem as I and these other working-class students did, what must she have thought about the editors' claim that she and the city college students had been blinded by cultural myths of the working-class?

Even Readers of Popular Mechanics

After thinking about this first concrete example of the difference between critical and non-critical reading, Louise may have noted "The Myth of Individual Opportunity," the title of the first chapter, which undermined her husband's struggle to make it through night school and her dream of becoming a nurse. For the first assignment, Stanforth had asked the students to read Arlene Skolnick's "The Paradox of Perfection," the lead essay for the chapter, "Harmony at Home: The Myth of the Model Family." Stanforth warned them this would be a difficult essay to read—and it was.

In spite of her negative reaction to the introduction of the book, Louise may have looked forward to reading about "Family." For the last fifteen years, Louise had devoted her life to the construction of a tight-knit family, herself, her husband, and their three children. To her, family was "the foundation for success in everything," she said (Durst 1999, 132). But I can imagine that the first sentence of the essay made her angry: "The American Family, as even readers of *Popular Mechanics* must know by now, is in what Sean O'Casey would have called 'a terrible state of chassis'" (Skolnick 1992, 403).

Louise may have thought of her husband, a reader of *Popular Mechanics* and wondered: "*even* readers?" She wouldn't have to conduct a sophisticated textual analysis to hear the condescending tone of the writer toward readers like her husband and several other male members of her extended family, the tone of the middle-class academic toward people who work with their hands, the tone that makes members of the working-class, like my father, contemptuous of people who can't *even* change their own oil.

Reading this essay, Louise would have to break though an affective social class barrier signaled by "*even* a reader of *Popular Mechanics*" [my italics]. She would have to accept the tone of distance that mocks people who do things instrumentally with their hands, the academic stance that blocks readers like her and her husband with a reference to Sean O'Casey. Louise had probably never read Sean O'Casey, so she has no idea of the triple layer of meaning in the predicate, "is in what Sean O'Casey would have called 'a terrible state of chassis.'" As academics, we read Skolnick's signal about the drunken working-class father and disintegrating families, and the secondary signal that Skolnick possesses reams of cultural capital and what kind of reader she expects. We can perhaps read more deeply and catch the sign of who is not allowed into her text. Louise is excluded because she would have to struggle to figure out the meaning of that first sentence while ignoring the writer's stance toward working-class people. With a little luck, Louise might guess that Skolnick means the family is in a state of crisis, a claim Louise would angrily contest.

The next sentence doesn't give Louise much help. It begins "Yet, . . ." Yet? She must have wondered, why "yet"? "Yet there are certain ironies" (Skolnick 1992, 403). Louise may not have understood "ironies." This word was probably her enemy, as it was mine until later than I care to admit. She may have had heard her voc-ed track (where she had been placed in high school fifteen years ago) English teachers use it—"irony," "ironical"—but the meaning was veiled. The word is a class signifier. It signals standing back, an objectification that allows the speaker to give "other" meaning to the signifier. Bourdieu (1984) claims the ironic stance is in fact the signal, par excellence, of the elite, who can afford to take a distanced stance toward material reality (34). To Louise and her working-class community, saying other than what one means is frowned upon. Working-class people make a virtue out of saying things straight (Bourdieu 176-177; 199-200). That verbal play of "irony" might occur in bars, locations outside the serious work of the day, but even there it would not be called "irony." It would be called a "put-on," or perhaps "bullshit," as in "you're bullshitting me."

Skolnick's sentence continues: ". . . ironies about the much-publicized crisis [ah, it was "crisis"] that give one pause" (403). Without being able to make any sense of this first paragraph, a jumble of nearly empty signifiers, Louise might intuit the studied tone of "much-publicized crisis" and "gives one pause," the kind of phrasing she would never hear in her everyday speech situations. If she heard someone say something like that, she would think the speaker pretentious. One imagines a speaker in the bar Lindquist's (2002) recreates: "It gives you pause, does it?"

Even a casual reading of the rest of the first page shows the kind of trouble Louise was in. As Gee (1998) argued in an analysis of a sentence about Bell's theorem (xi), there are in this theoretically explicit text logical gaps between paragraphs and sentences that the academic reader would know how to fill in, requiring a certain suspension of meaning, reading strategies of glancing ahead to find the main verb and reading back to make connections; inclusions of references with in-text citations that tell Louise nothing; uses of academic conventions ("e.g.") that Louise, although she has seen them, doesn't know yet; obscure phrases, ("mass-produced images [that] have 'extraordinary powers to determine our demands upon reality'"; "the Family is vulnerable to confusion between truth and illusion"; "we know The Family, in the aggregate, only vicariously." What? What?

Durst (1999) tells us that Louise struggled through this essay three times to make sense of it; even then the best she could do was grab onto some paragraphs where the writer seemed to say something with which Louise could agree. But she did understand that the writer's attitude toward the family as an institution was negative, and she didn't like it. She flipped through the other five essays in the unit to read the last one, "The Gay Family" by Richard Goldstein. This was going to be assigned later, but she decided to read it after Skolnick's, maybe to find out whether the class was going to be like this for the rest of the semester.

Goldstein's essay was easier to understand, but it offended her sense of family—what family is and what it means. The essay was written by a homosexual who had as his clear purpose the legitimization of homosexual families, two men or two women raising children. The writer wanted the reader to understand homosexual families as alternative family structures. There were descriptions of homosexual weddings—guitars, same-sex partners openly kissing.

Louise might have tried to be broad-minded, but her working-class community did not accept this kind of alternative family structure. Many people she knew were intense in their opposition to alternative structures for anything. They believed in God, and to Louise, God is pretty clear about His (there's little possibility of a Her) position on the issue of alternative family structures. Louise might try not to judge homosexuals, but she would be certain that their sexual orientation, much less the notion that they should be able to raise children, was a deviation from what was *meant* to be. She might also think that she shouldn't have to read this kind of essay in a class that was supposed to teach her something about writing.

Academics may be intolerant of the working-class opposition to alternative family structures, but the opposition might more usefully be understood as the consequence of the dialectic between the working-class experience and habitus. Both Bernstein (1971) and Bourdieu (1984) make clear that members of the working-classes are born into a social structure with a limited variety of available roles—or as Bourdieu describes it, life-chances. Working-class people learn how to fit within the roles rather than adapt available roles to their own inclinations. Bourdieu argues this kind of habitus is another example of the dialectic between material conditions and habitus: people are inclined to re-enact the actions of the stories that are told about them, thus contributing to the continuation of the stories. Working-class people's limited roles bracket the story in the home and community. Similarly, in the workplace, working-class people have to adapt to material conditions rather than adapt the material conditions to their desires. Consequently, working-class people value predictable and traditional gender roles. Men are supposed to be men and women, women. This differentiation is a function of labor, the primary working-class commodity. Labor means physical work, which in turn means muscle, masculinity, the determinant of what it means in the working-class world to be a man. The woman, in turn, should exhibit features and behavior connected with childbearing and running a home. Upper-class men, by contrast, cultivate an effeminate look (Bourdieu 171, 190-191) and upper-class women try to look as if childbearing might have been an accident that didn't really happen. In the working-class world, men and women are supposed to look and act differently; in the upper classes, they cultivate the same look (382).

The decreased gender differentiation in the upper classes leads to more ambiguous gender identities. Bourdieu (1984) argues that the increasing tolerance of deviations from sexual "norms" is directly related to upper-class men being removed from labor and, in effect, from necessity. Muscle/maleness is no longer a useful attribute; in fact, it is negatively interpreted as a working-class feature. In place of muscle, the upper-class male values style—a distinctly effeminate concern for the working-class male. Because of his historic work situation, the working-class male is more inclined than a member of the dominant class to reject effeminate males.[1] This rejection is generalized to a rejection of the homosexual life style, predicting Louise's negative response to the essay on gay families. Because they are socialized in such a different habitus, middle-class academics might misread the working-class "structuring structures" that organize explicit gender roles. They might interpret the concomitant rejection of homosexuality as intolerance. While imagining themselves as critical thinkers, they could miss that like working-class people, they are mostly living out the story that was told about them.

In addition to rejecting alternative family structures, Louise adhered to the working-class sense of community. The working-class notion of community is local, focusing on the immediate family, the extended family, and the local community of a neighborhood or small town (see Lareau 2003). "Community" does not mean a large urban or national affinity group, simply because the social logic prevents working-class people from participating in these larger communities. The local sense of community is connected to the limited roles available to members of a small community and the working-class socialization into a position-oriented habitus. When people submerge their personal identities within the need of the local community, the community functions smoothly and thus becomes a social force that protects the members from the outside. The social logic of community is accentuated, for example, in

1. In Bourdieu's (1984) survey, 16% of manual laborers said they personally knew homosexuals compared to 37% in the middle classes, (382).

unions. The importance of adapting to the needs of the community and socialization into a position-oriented habitus works to maintain community and working-class solidarity, their fundamental weapon against the capitalist exploitation of labor.

Consequently, the first pair of essays that Louise read in English 102 was a triple assault on her literacy and ethos. The language excluded her; her religion of "family" was criticized; and alternative family structures were promoted. When she went to class the next day, Louise had a good idea of Stanforth's point of view on these issues. Durst (1999) says that as a consequence of the readings, Louise began the quarter with an "us versus them" attitude, with "them" being Stanforth, the essayists, and to a certain extent, the liberal university establishment within which a required writing course was the cover for a left-wing attempt to undermine her most deeply held beliefs (131).

DIFFERENCE AND DIVERSITY

Stanforth's course is typical of writing courses that promote the social strand of critical thinking. Difference and diversity are key concepts in these courses, reflecting the teacher's hopes of promoting a tolerance of difference and encouraging diversity. Difference and diversity are twin themes in *Rereading America*, perhaps accounting for its popularity among teachers and by some perverse logic for its equal infamy among students. The teachers' attitudes are signaled by its Amazon sales rank, 736. I looked at three other readers: *A Reader for College Writers*, ranked 33,407, and *Patterns for a Purpose*, ranked 26,136, and *Ways of Reading*, ranked 3,466. One might also compare *Rereading America*'s rank to the arguably most popular rhetoric, the *St. Martin's Guide to Writing*, ranking 713. Among other rhetorics, *Everything's an Argument* ranks 2,424, and *The Prentice Hall Guide to Writing* ranks 6,699. In spite of its popularity among teachers, Durst (1999) says *Rereading America* was universally disliked by the students in English 102 (157). Students' comments on Amazon's reviews reflect this student resistance to the text. Shellie (2006) writes:

> I took a class and was required to read this book. It was the worst book I have ever HAD to read! I struggled to finish the reading assignments. This book was so biased I couldn't believe it. Once again America is the root of all evil, and the white man is the devil. The authors need to take another look at America, and see that there is some good going on here! Only someone that is so miserable and bitter would believe (or write) this book. They had

> nothing good to say about America. . . . [I]f it is so bad here, why don't they move (and take along anyone that agrees with them).

By the number of books she reads and reviews, Shellie looks like an avid reader and reviewer, the kind of student we need to take seriously rather than the kind of conservative griper a teacher might assume from her comment. Nevadagrl435 (2004), another avid reader and reviewer writes, "This book is pure mindless drivel for the uneducated mind! *Rereading America* is chock full of liberal antidotes on our society-from education to class. Read this book and you will find that it feels that college freshmen and sophomores are sheltered conservative creatures who need their eyes opened." Interestingly, she adds at the end of what could fairly be called a diatribe, "The two star rating is for the articles by Moore and Ehrienrich, the only two that I could stand out of this entire book."

It looks as if two students may have found the book useful, although one of the student comments seems to be from a teacher who was taking a graduate course in ESL (Corzo 2000); thus, her comment positions her in a hybrid situation. Without exception, however, the reviewers clearly identified as teachers attest to its value. "Freshmen cannot," Jennifer Smith (2007) writes, "literally, think outside the box. They need readings like these to open their eyes and begin to think critically." Jean Rhys (2007) writes in response to the student complaints about political indoctrination, "I don't think this text aims to change anyone's political or social views from right to left, although it does provoke in-depth critical thinking and writing that should be at the center of any first-year curriculum at the college level."

The differences in these reviews dramatize the contradiction, as Ellsworth (1989) puts it, "between the emancipatory project of critical pedagogy and the hierarchical relation between teachers and students" (308), the teachers presumably "outside the box" and the students trapped in it. The irony is doubled if one recalls Freire's description of "certain members of the oppressor class" who

> join the oppressed in their struggle for liberation, thus moving from one pole of the contradiction [i.e., between oppressed and oppressor] to the other. Theirs is a fundamental role, and has been so throughout the history of this struggle. It happens, however, that as they cease to be exploiters or indifferent spectators or simply the heirs of exploitation and move to the side of the exploited, they almost always bring with them the marks of their origin: their

> prejudices and their deformations, which include a lack of confidence in the people's ability to think, to want, and to know. (42)

Although Freire is here referring to sociopolitical oppression, we can reframe his analysis within the classroom. The narrative of student resistance to critical teaching frequently features teachers who imagine themselves on the side of the oppressed, and by extension, on the side of students in the student/teacher contradiction. But the teachers ineluctably bring to the classroom situation their experiences of having been transformed from student into teacher, experiences that carry with them gestalts that invade their minds, transferring their allegiance to the other pole of the contradiction. The teachers struggle in their desire to resolve the contradiction to engage their students in the larger struggle between the dominated and dominating social groups in our culture, aligning, as Jerry Farber (1970) notably did with *The Student as Nigger*, the larger social contradiction with the more restricted one in the classroom. But many students refuse to buy this analogy. They see the teachers as threatening them, primarily with grades. The critical teachers may be inclined to externalize their failure and blame the students for having been trained to think "in boxes" and they continue to think their job as teachers is to teach students how to think critically—as if the student/teacher opposition were paired with naïve/critical thinking.

I witnessed a telling example of this contradiction in a conference presentation at a Pedagogy and Theatre of the Oppressed Conference in 2001, a conference ironically focusing on the works of Freire and Augusto Boal. Like Stanforth, the teacher presenting this paper was a socially conscious, committed teacher. Nevertheless, his presentation made clear that by trying to promote difference and diversity, he was alienating his students with the effect of further disempowering the students he had meant to empower. I relate this incident with full respect for the teacher, Gary Cale, for he was in his presentation holding his example up to critique in a serious effort to ask of his colleagues what had gone wrong.

The students in Cale's (2001) first-year writing class were non-traditional students, four African-Americans and thirteen Anglos—most of them women. Cale's agenda was to help these primarily working-class students to "see anew." Cale's course "emphasized oppression and privilege and democratic teaching practices such as a negotiated curriculum, direct challenges to hegemonic culture, and bringing to voice marginalized students." The semester was divided into four units (social constructionism, classism, racism, sexism) with students developing "at the end of each unit . . . individual action plans to combat the oppression they noticed in their own lives, as well as to write a short paper on some aspect of the issue being studied." The grammatical functions in this description of "a short paper" and "actions plans" indicate their relative status—the "action plans" are clearly primary and the paper is secondary. Since the short paper was developed at the end of each unit, we can infer that writing has been largely displaced by the reading and discussion that constituted the bulk of the course. The rest of the conference paper suggests as much: the bulk of the presentation was devoted to the student/teacher conflict that was the consequence of Cale's transformative agenda.

The traditional argumentation model predicated on students debating each other turned toward the teacher debating the students in an increasingly serious confrontational pedagogy. Cale interpreted the student resistance in terms of Strickland's (1990) hegemonically induced will-not-to-know. Cale's purpose of getting the students to "see anew" really meant getting students to see reality from his postmodernist, neomarxist point of view, resulting in a face-off, which Cale describes like this:

> Although in class discussions, readings, and lectures, I emphasized the institutional and symbolic/cultural levels of oppression and privilege, many adult learners remained committed to the position that racism in particular, but also sexism and classism, is an individual phenomenon. They consistently resisted the concept of asymmetrical power relations and denied that the dominant culture exerted any control, let alone hegemonic control over their lives. Perhaps more significantly, they argued that oppressed people of color could be racists, regardless of the clear power differences. Despite the many class periods spent on discussions and activities that attempted to make visible the many embedded racist and classist discourses, symbols, representations, and institutions that help to form our worldviews. . . . They expressed surprise and dismay that I insisted a problem existed, telling me to "get over it."

Although Cale (2001) claimed that one of his two purposes was "to disrupt traditional hierarchical power relationships between teachers and students through the use of radical pedagogy," he reifies the student/teacher contradiction. Because of his serious commitment to the larger project of social reform, Cale can't see that his pedagogy is working against its announced purpose. It almost seems as if there is an unacknowledged transcript that ranges teachers and students against each other, no matter how fervently teachers and students want to collaborate. We can see this transcript working when Cale explains that at "various points during the semester my desire for creating and nurturing a democratic classroom were upset by student resistance of various kinds." The structure of his sentence points toward his phenomenological interpretation of the classroom drama: his "desire for creating and nurturing a democratic classroom" is the noun phrase as if it were existentially present as an object to be accepted, rejected, or destroyed. Agency is deferred by the passive construction to the student resistance—the *resistance* overturns his desire for a democratic classroom and by inference for a more egalitarian social structure.

In an attempt to rescue his project, Cale resorts to the fundamental colonialist strategy that Freire ([1970] 1995) describes as "Divide and Rule," an oppressive strategy "as old as oppression itself" (122). Cale tries to enlist some students on his side so that the "discussions" wouldn't seem like the radical teacher facing off against the students. Here's how Cale explains it:

> On several occasions White students waited for me after class or chatted with me on breaks about how upset they were by comments made by other White students during class discussions. . . . but they did not offer these critiques as part of the public discussions on racism. This was especially disturbing to me because I repeatedly asked for their assistance, requesting that they voice their opinions during our plenary discussions and act as allies for the students of color.

"Students of color" is a slant phrase, pointing at the four black students, who seem to be angry about having been made the object of discussion, but also self-referentially at Cale *through* the black students. Cale (2001) struggles to align the four blacks and a few whites with him in his attack on the positions held by the working-class students, who in impolite liberal circles might have been called rednecks. But his factionalizing strategy fails. The hoped-for "students of color" allies wouldn't enter the discussions; they skipped classes; they dropped the course. Even the after-class group (Shor 1996, 161-63) failed when the members "chose not to call . . . or email" him although he had specifically asked for their votes on key decisions. "They simply avoided me," he writes. I imagine their eyes looking the other way.

I have wanted to look the other way as well, because I am uncomfortable analyzing Cale's (2001) interpretation of classroom events. By publicizing his interpretation, Cale has put himself at risk, as we all do when we write. My purpose in analyzing Cale's story is to locate cause, not blame. I want to move beyond an interpretation of this story at the personal level to the structural level. At this level, I suspect that Cale has inadvertently acted as an agent for the dominant portion of the social structure. He ends up blocking not only his students' intellectual development but also their chances for gaining some of the privileges that the social structure denies them. They came into Cale's classroom because the first-year composition course was required. Many of them may have had instrumental motivations, like Louise, of learning something about writing that would help them in their other classes and careers. But what they got was a confrontation in which their values were placed in opposition to the teacher's and against one another's through Cale's attempt to factionalize. The end result was a class in which over a third of the working-class students dropped the course. In his struggle to get students to think critically about social structures, Cale impeded these students' attempts to move across social classes by improving their abilities to write—a key to social mobility for those who lack social and cultural capital.

ALLODOXIA: PUTTING THE BLAME ELSEWHERE

Bourdieu (1984) describes *allodoxia* as an important strategy of social reproduction. Events determined by asymmetrical social relationships create opportunities for misreading by those who are furthest from the centers of power (142). They think they are seeing one thing, but they are seeing another, as if they were playing a shell game with the dominant fraction being a trickster who with his little finger flips the pea to the thimble on the right while the mark's eyes remain on the thimble to the left. Bourdieu describes as an instance of allodoxia the refusal of graduates to acknowledge the real as opposed to the nominal value of their college degrees. Working-class students who earn degrees in less valued professions (like English and philosophy) from unranked universities are surprised, for example, when they can't get good jobs after graduating. Rather than understanding the hystereisis effect consistent with the proliferation of undergraduate degrees (see above, 27), these students may put the blame elsewhere, for example white males might angrily cite affirmative action policies as the reason for their inabilities to get good jobs (Durst 1999, 142). Working-class students might similarly put the blame for their low grades in writing classes on their teachers when it might more accurately be located in social structures that privilege language and extrinsic knowledge they don't have, or more subtly, in classroom activities designed to help them but that in fact push them further from the center of power (Clark 1960; Shor 1992, 109).

The Rhetorical Situation

In their collaborative report on "What Happened in English 101?," Scott Hendrix, Sarah Hoskinson, Erika Jacobson, and Saira Sufi (2000) describe a notable example of allodoxia, the students and graduate student teachers, all far from the centers of power, misreading each other, the consequence of which was students who should have learned more about writing than they did.[2]

As with Stanforth's class (Durst 1999), the working-class students in English 101 misread the teachers—Hendrix and Jacobson—from the start, the consequence of a quotation from Janet Zandy that Hendrix and Jacobs had included in the syllabus. Zandy had written: "Let us imagine what it would be like if the history and culture of working-class people were at the center of educational practices. What would students learn?" (Hendrix et al. 2000, 54). Zandy had one thing in mind—interrogating by imagining its opposite the hegemonic privileging of middle-class language and culture in academic settings (and the concomitant silencing of working-class language and culture), but the working-class students, as Hendrix and Jacobson belatedly came to understand, had quite another. Rather than imagining the educational project from an unfamiliar point of view, the students thought Hendrix and Jacobson were implying through Zandy's quote that it's a good thing institutions don't base their educational practices on working-class language and

2. Although I am critical of this class as Scott Hendrix and Erika Jacobson report it, my critique should not in any way be directed at Scott and Erika, both of whom I know and respect. I know that this experience was a learning experience for Scott, who is now a writing center director at Albion College—and who I in fact first met when he was a graduate student and attended one of our early Pedagogy and Theatre of the Oppressed conferences at Omaha, Nebraska. I have communicated by email with Erika, who to my disappointment, decided not to continue with her graduate degree. I think she would have been a wonderful teacher. I worry that this experience may have discouraged her from continuing her studies.

culture because if they did, students wouldn't learn anything worthwhile. As a consequence of misreading the teachers' reason for including the quote, the working-class students misread the teachers' attitudes toward the students' working-class culture.

People forget how they read the world from the other side of an extended series of transformative experiences. Professors forget their graduate student perspectives, graduate students forget when they were first-year students, and tenured working-class academics forget when they were really working-class. People on different sides of these social locations look at the same set of words and read different meanings. Thus, Stanforth (Durst 1999) missed how Louise may have read the dismissive "*even* readers of Popular Mechanics" (my italics), and Hendrix and Jacobson, missed how their working-class students would read Zandy's comment.

Students similarly misread Hendrix's (Hendrix et al. 2000) remarks about the way the deck was stacked against working-class students. The students came to know this repeated conversation as the "weeding out" factor. Hendrix wanted them to understand how the dominant classes infuse hegemonic practices into a supposedly neutral institution, but the students interpreted his comments as both a threat and a promise: that if they came from the working-classes, there wasn't much chance for them to succeed in this class; if Hendrix and Jacobson didn't weed them out, later professors would.

Sufi, an empowered student who ultimately organized a group of students to complain to the writing program administrator about the politicization of the class, thought Hendrix in particular seemed to be using the "weeding out" factor as a threat. She wrote that Hendrix thought the students "just accepted that English 101 [was] a 'weed out' class. It seemed as if he were saying we were not intelligent enough to do anything about it" (Hendrix et al. 2000, 59). Hoskinson, another student, wrote that Hendrix made it clear "that some (or all) of us were simply fated to fail regardless of how hard we worked, what we learned, or what we planned to do" (61).

The effect of Hendrix's attempt to unveil the mechanisms of social reproduction for the students was to take away the hope with which they had entered the class. Jacobson noted that

> [t]hroughout the semester different students would bring the discussion back to "weeding out" and I never listened long enough to hear the personal

> frustration and anger. Only much later did I realize what I wasn't hearing. Students were continually asking Scott and me for more explanation of terms like *hegemony* and *ideology*. Their questions were surely a form of protesting our heavy-handed use of politically volatile academic jargon—teacherly jargon that seemed designed to keep students at an arm's length from their goals of success at the university. . . Drowning in talk of *hegemony* and *ideology*, students may have already felt negated: weeded out. (Hendrix et al. 2000, 61)

Although this message of hopelessness was unintended, Hendrix's underlying pessimism may have been working its way through into his classroom discourse. He writes about the $7.50 an hour telemarketing jobs he felt many students were destined for: "But it frustrated me that for too many of my students (and others not my students), this would be their best job offer even after they've completed their degree, in our downsizing, de-skilling economy, flooded with growing temporary positions and fewer and fewer living wage, benefit bearing, worker-respected jobs" (Hendrix et al. 2000, 62).

Hendrix's reflections echo the dialogue between James Berlin and Stephen North that appeared in one of the "Transcript" sections of C. Mark Hurlbert and Michael Blitz's *Composition and Resistance* (1991). Speaking of the underlying purposes of required writing courses, Berlin says: "You know what? I don't think I'm preparing them for the really good life. I mean they're going to have breakdowns. . . . I mean, it's not really that good a future anyhow, if they're going to be unhappy, right?"

After some protest from North, Berlin continues:

> *Berlin:* I want them to know what's in store for them! I mean, they have narratives—
>
> *North:* You're like an image broker. Here's a picture of poor people, here's a picture of unhappy rich people.
>
> *Berlin:* No, I want to show them a picture of themselves in a few years and how unhappy they're going to be. (Hurlbert and Blitz 1991, 137)

This dialogue doesn't capture the essence of Berlin—I think it caught him in an unedited moment in which he displayed a dark image of subjective reality, as he saw it within that particular conversation; nevertheless, if you overlay Hendrix's and Berlin's visions here, they are indistinguishable. And if you overlay the purpose of English 101 and the experimental courses Berlin popularized at Purdue, they would likewise be indistinguishable. Berlin's purpose in those courses was to get students to read cultural structures as codes. English 101 had exactly the same purpose with a more particular focus on education—which really meant that the focus was reflexive in an unintended sense. Hendrix and Jacobson intended to aim the reflection at all courses but theirs, but the students applied the same decodifying principles to the course in which they were at that point submerged, "drowning," as Jacobson said, "in talk of *hegemony* and *ideology*" (61).

The Writing Situation

In their desire to encourage their students to think critically about education and culture, Hendrix and Jacobson (Hendrix et al. 2000) framed their writing task to point students toward ways of thinking but seem to have given them little guidance on how to write, as if the emphasis on thinking critically about culture had obscured writing instruction. Here was the first writing task: after the students had read a short story ("Push" by Sapphire), Hendrix and Jacobson asked the students to "determine for yourself—and then detail in your writing and for your audience—the key element of education in the short story." They were then to "pick a single key educational element in the story and build your paper around this idea" (Hendrix et al. 2000, 55).

The writing task, as Hendrix and Jacobson described it, offered no clues about the specifics of a rhetorical situation for the writing task or what genre would answer the demands of that situation. We see no evidence in the article of Hendrix and Jacobson presenting samples and rhetorical analyses of other responses to similar kinds of writing tasks in similar kinds of rhetorical situations to analyze as models. As well, the key phrases in the assignment are sliding signifiers that confused the students (e.g., "detail in your writing," "key educational element"). The teachers had hoped the students would think critically about the story, connect their own lives (significantly different from the main characters) with the life of an African American girl in Harlem who manages with the help of a teacher (one of Miller's hero-teachers) to "push" herself into literacy. What they got were impersonal plot summaries (Hendrix et al. 2000, 55-56).

For the second writing assignment, Hendrix and Jacobson asked the students to "Read Rose [*Lives on the Boundary*] in light of Ede [the required course rhetoric], Sapphire, and your own schooling experiences—to put together the pieces of the texts and stories we've read so far in this course—and to determine how and why Rose complicates and/or extends Ede's discussion of writing as a 'process'" (Hendrix et al. 2000, 56). Again, this is a kind of genreless writing task within an empty rhetorical situation, pivoting on vague, or to the student meaningless, verbs ("determine," "complicate," "extends"). The "[p]rompts were vague and unreadable," wrote one student in the course evaluations (Hendrix et al. 2000, 56). The consequence, once again, was bad writing.

Bad writing—or at least writing that fails to meet a teacher's expectations—leads to bad grades, which the students may have misread as proof that the instructors were following through on the misperceived threat that working-class students would be weeded out. Here was social class reproduction, the theme of the course, in action. Hendrix and Jacobson of course had no such intention; they were signaling by bad grades the distance between the essays they had in their heads and what they were receiving from their students. Hendrix and Jacobson wanted essays that would make "connections between [the students'] readings and their experience, . . . complicate their views, . . . grapple with the educational myths that the texts challenged." Overall, they claimed, the "papers lacked critical thinking" (Hendrix et al. 2000, 56). By the low grades, Hendrix and Jacobson were surely signaling how far, in their view, the students had yet to go, not that they couldn't get there.

"What Happened in English 101?" is filled with other instances of misreading, the students misreading the teachers' comments, the teachers' misreading the students' questions, perhaps all instances of allodoxia hinging on the teachers' and students' different purposes. The students with their instrumental purposes wanted to learn about writing and ways to get good grades; the teachers wanted the students to see through the cultural myths that had shaped them, to see the world and themselves anew, to become critical citizens of a better world. Writing was the wave that would carry this new way of seeing. The result was students who didn't learn about writing and teachers who, as Hendrix put it, were "hammered . . . in semester-end evaluations" (Hendrix et al. 2000, 59).

David Seitz (2004) describes an example on his comparable misreading of a working-class student's text. But whereas the student/teacher and teacher/student misreading in "What Happened in English 101?" seemed to be a function of their different motivations, Seitz adds to the discussion his social location in the intellectual fraction of the middle-class and his misreading of his student's social location. Seitz warns that while critical teachers often accuse students of naïve, unreflexive reading, critical teachers too easily forget that they, too, are reading students' texts from a position to which the teachers are chained as much, perhaps, as students are chained to theirs.

Seitz (2004) had asked his students to compare educational experiences in informal and formal educational settings, drawing on conceptions of "connected learning" from *Women's Ways of Knowing* and Jean Anyon's "Social Class and the Hidden Curriculum of Work." Seitz doesn't give us specifics of the rhetorical situation, real or imaginary, of the task, but he makes clear that the purpose is for the writers to connect conceptions from these two seminal works with the students' contrastive experiences in order to illuminate something about the interrelationships between social class, educational institutions, and social structures. His purpose does not seem to be grounded in any specific set of writing skills but rather in promoting a critical interpretation of educational institutions, and by extension, of other social institutions.

Seitz's (2004) working-class student, George, wrote in response a somewhat clumsy essay, made clumsier by his attempt to appropriate the teacher/reading's terminology, like "connected" and "communicative" education in his analysis of his experiences. In his first response to George's essay, Seitz was critical of George for misusing terms, the consequence of which was an uncritical approval of values that Seitz thought should be interrogated. George turned concepts used in the readings to challenge traditional social structures into instrumental tools for gaining economic and social advantage. In his essay, George explains that by learning to use his words well, he will be able to fool others into buying shoes they don't need, an application of "connected education" that would have made the authors of *Women's Ways of Knowing* wonder where they had gone wrong.

Two issues are interesting in Seitz's (2004) discussion of his interpretation of George's essay. I am struck first by the content of his discussion: Seitz's reflection on his responses concern only the ideas George discussed in his essay. Seitz says, "my margin comments to George gave him openings to argue against the readings' assumed values of connected education based on his experiences" and "I am arrested now by the less open questions I posed—the questions that implied a more politically correct agenda" (13). Although I realize that one separates writing and thinking at one's peril, I nevertheless notice that Seitz is not telling us anything about his suggestions to George on how to improve his writing, his word choice, his sentence structure, his organizational strategies, or how he met genre-specific criteria. As with so many responses one reads about in articles by critical teachers, Seitz's suggestions are exclusively concerned with George's thinking.

Secondly, I am impressed by Seitz's (2004) linkage of social class location with his interpretation of and response to George's essay. Because of George's content in the essay, comments on his Oak Park high school, and his linguistic competence, Seitz had assumed George was the prototypical middle-class male going to college only to earn a degree that would secure his own middle-class position; that is, George is the student about whom critical teachers like Strickland (1990), Berlin (1991), and Knoblauch (1991) complain.

Only later did Seitz (2004) realize that George was located in a different social space and was consequently tracking a different trajectory, angling up rather than staying flat. This realization made Seitz acknowledge his arrogance in presuming from his middle-class location to challenge George's instrumental purposes. It was this reflection, in fact, that impelled Seitz to write his book in a search for more productive ways of helping students learn how to think critically about social structures and assumed knowledge.

The Essay

For a last example of misreading, I will return to Stanforth's and Durst's (Durst 1999) interpretation of one of Louise's essays. As I have been with Seitz's, I am impressed with Durst's ability to interpret the claims of critical pedagogy from the working-class students' perspectives that calls into question pedagogies based on Berlin's project of teaching writing through the lens of cultural critique. Nevertheless, I think that with the best of intentions, Durst misreads Louise's essay and the cause of her failure. This kind of misreading, based on an inflation of argumentation as the central academic supra-genre, disproportionately disadvantages working-class students who have not been trained as well as middle-class students to play that particular game. This misreading also blocks an alternative mode of instruction that could have helped Louise improve her writing rather than confirm her failure.

Multivocality is a distinguishing feature of both critical literacy and academic writing. Students need to learn how to investigate a range of positions on any issue and give alternative world views a fair hearing, suspending judgments based on the values they learned in their home communities. Multivocality is the bedrock of the academic frame of mind—being inside a conversation and fully representing that insideness by having one's own voice intermingle as if in a cocktail conversation with other voices and perspectives. If there was one criteria in Stanforth's (Durst 1999) class for the successful essay, multivocality was it. If there was one characteristic of failure, it was bringing in someone else's voice as if it were a hot potato.

Stanforth asked her students to consider the essays they had read on the subject of diversity and prejudice, focusing in particular on an essay by Gordon Allport about the formation of in-groups, the marginalization of out-groups, and the appropriation of reference-groups—an outside group on which the in-group members model their behavior. Stanforth asked the students to write narratives about their own membership in groups but to reflect on their narratives by considering what the other writers (i.e., Allport) had said about group behavior (Durst 1999, 152-54).

Louise wrote a particularly unsuccessful essay in response to the writing task. Durst (1999) faults Louise for her inability to understand the assigned readings and argumentation as a genre (130), but I want to analyze Louise's failure as the product of a teaching situation that veered off-track after colliding with Stanforth's critical literacy agenda.

Although Stanforth was a talented teacher,[3] particularly sensitive to the needs of students who belong to marginalized socioeconomic groups, the politics of her course and the writing task combined to make Louise, a non-traditional, working-class woman, look like a worse writer than she was. Consequently, Stanforth had to respond to her in a more negative fashion than if Louise had been placed in a writing situation that would have allowed her writing and thinking skills to bloom. One might go further and say that the concrete effect for Louise was a lower grade than she might otherwise have earned. This lower grade, while not terminal, could have consequences for the rest of Louise's academic career as well as for her image of herself as a writer.

Although the readings and Stanforth's obvious orientation were pointing toward race, gender, or class as the preferred topics for the essay (with race being the favored of the three), Louise, like most of the other students, avoided these controversial issues. Louise chose instead to write about her membership in two church groups as the subject of her "in/

3. After a break, Stanforth finished her PhD and began teaching at Thomas More College where in 2004 she was named Teacher of the Year. From personal communications I have had with her and from Durst's report, I would say she is an outstanding, student-centered teacher.

out group" essay. Writing teachers know that a student's choice of topic constrains his or her chance for success. In the choice of her topic alone, Louise was going against the odds—particularly in a critical literacy class. Not only was Louise going to write about her religion, but she was going to write about her membership in a Baptist church. To most critical teachers (myself included), displaying one's devotion to a fundamentalist religion is like advertising one's inability to think critically. In addition, membership in the Baptist church, particularly in the North, signifies one's working-class status. One need only consider the difference in effect if Louise had chosen as her subject her membership in the Unitarian church.

Stanforth, who came from a working-class, Appalachian family, probably had more tolerance for Louise's religious orientation than Strickland (1990) seems to have had for his fundamentalist students. But we must also remember that Stanforth was in some sense escaping her background; she was becoming a member of the academic class, one feature of which is a cynical attitude to stories about immaculate conceptions and saviors raining down from the sky.

Louise missed the "right" topic—she had considered writing about "growing up in a working-class neighborhood and feeling somewhat inferior to her classmates who came from a more middle-class neighborhood and tended to look down on her and her peers" (Durst 1999, 154). Instead of picking a topic that would have given her the chance to explore the iniquities of group membership and excluding behavior, which the readings invited, she chose a topic that oriented her primarily toward the positive effects of group membership. She had a chance to hit a home run, but she bunted.

Here's the opening of her essay, "11:00 A.M. Sunday Morning":

> "C'mon Mike. Let's go we're gonna be late." I yelled as I paced up and down the room looking for my purse. "I'm coming," he snapped as he entered the room. It was Sunday morning and I couldn't wait to get to church. Our thirty-minute drive always seemed like three hours. (Durst 1999, 155)

From a working-class woman who had been relegated to the vocational track in her high school and had been fifteen years out of school, this is an interesting opening. I see strong composing instincts here. She has learned to open with the dramatic moment. Some writing sophisticates will challenge the dramatic moment as a threadbare strategy, but my point is that she has *learned* this as had the eighth-grade student Paul (1993) criticized (see above, 58).

She has also learned how to catch the speech pattern: "C'mon," "gonna be." She has learned to use descriptive verbs: "yelled," "paced," "snapped." In what is essentially two sentences, she has not only set a scene but also hinted at different attitudes between her and her husband toward the activity that is going to be the subject of her essay. She then moves out with some skill to the traditional panoramic shot: "It was Sunday morning and I couldn't wait . . ." Panning out to set the context is also a skill she has learned. Louise is the kind of writer I like to work with. In a few minutes, we could show her about direct address, run-ons (which in this case was mimetically effective but still should not have been used), proper punctuation with tags, and misplaced modifiers.

In her next paragraph, she pans further back to describe a broader context for the subject of her essay—feeling like an insider with one church (good) and an outsider in another (ambivalent). She describes the "Longview Baptist Church," the "love" and "warmth" the members felt for each other in spite of their "diversity." She then describes the pastor, who is clearly the fundamental reason for the "one big happy family" in spite of the "diversity." She concludes her description, "When he was on a one on one bases with me he could make me feel like I was such a unique person. It was like I was the missing piece to the puzzle that was needed to make it complete" (Durst 1999, 155).

In this section, she has made her moves to satisfy Stanforth's requirements—mixing these strategies almost seamlessly with her more personal motivations for writing, which were to work out through writing her decision to switch churches. She includes a reference to "diversity," one of the major themes of this unit. In addition, she has linked the concept of diversity with her stronger motivation for writing by interpreting the members of the Longview Baptist Church as a family. These links connect to Allport's minor thesis of the rare in-groups that structure their own meaning without devaluing outsiders. Perhaps more importantly, Louise shows her strength as a writer when she uses the missing piece of a puzzle image to explain how the pastor made her feel. A teacher could easily compliment her on some of these rhetorical strategies and also show her the kind of conventions to which she should pay attention, such as the importance of homonyms and the use of hyphens. (The misuse of "bases" for "basis" is fascinatingly close to misusing "chassis" for "crisis.")

In the next paragraph, Louise tries to respond explicitly to Stanforth's instructions: "I contribute my individual growth to the experience with my in-group. . . . Within my in-group there was a reference group. I would watch the people that were more mature in their Christianity than I was. They appeared to be wise and confident in their relationship with God. . . . I would think to myself, I want to be just like that" (Durst 1999, 155).

This break in what could have been an interesting essay is painful: Louise struggles to include something from the required reading. That struggle to please the teacher perhaps accounts for another malapropism as Louise reaches for "attribute" but comes up with "contribute." She can't quite fit Allport's concepts into the frame of her subject. Her references to the in-group and reference-group are only gestures, and she misses entirely Allport's definition of a reference-group as a desirable group *outside* the in-group. If she had chosen her original subject of her working-class status in a dominantly middle-class high school, she would have been able to interpret her different and conflicting social group memberships within Allport's schema. In essence, Louise made the mistake of writing about what she wanted to understand rather than writing about a subject that would have allowed her to meet her teacher's expectations. Louise's essay was a working-class response to a middle-class writing situation.

As Louise tries to accommodate the implied sociopolitical demands of the writing task, her writing self-destructs. She begins to discuss her new church, within which she considers herself an out-group member.

> Why is it I feel like I'm in the out-group? We have the same basic foundations of religious beliefs. Allport states that the in-groups preferences must be his preferences. It's enemies his enemies. One strong point in favor of this statement is that I don't' have the same feelings for the pastor as the in-group does. My opinion of him is not as favorable as some of the others, therefore I don't support him in quite the same way as the in-group does." (Durst 1999, 156)

I could have helped her with the first part of the essay, but the composing missteps here are overwhelming. In contrast to her auditory sensitivity in the opening paragraph, in this segment she isn't hearing her words. Her opening rhetorical question is a weak attempt to move toward an accommodation of the writing task requirements. She leaps from sentence to sentence, unaware of the transitional problems. She shifts her subject with each sentence: I → We→ Allport→It's. The real subjects of these sentences are "out-group"→ "beliefs"→ "preferences"→ "enemies"→ "feelings for the pastor." Readers can make the connections, but they have to work to make them. When Louise focuses on a "point in favor of this statement," she continues to throw her coherence off in part because she is losing her voice in an attempt to sound academic. She has in fact borrowed her phrasing directly from Allport (1953), who wrote, "A strong argument in favor of this view is . . ." (303), but as Bakhtin (1981) has pointed out in his theory of hybridization, words that seem natural to others seem alien when speakers first try them out (293-294). Having displaced her own language with her interpretation of academese, she contradicts herself and then loses control of meaning in her next sentence, which is close to drivel, the kind of repetition student writers engage in to fill out the required number of words with some vague hope that new information might come up as they repeat themselves.

It gets worse. In the next paragraph, she tries to accommodate what she thinks might be the academic model:

> Another point, is that I don't have the same feelings toward the members of the church as I did at Longview. At Lonview it was easy to bond with it's members, at Faith it's not so easy. The people are hard to get to know on a personal level. The members are just a different group with different characteristics. (Durst 1999, 156 [miscues in original])

You can feel the disconnect between Louise and her writing, particularly if you compare her prose here to the first sentences in her essay: "C'mon Mike. Lets go we're gonna be late." Toward the end of the essay, she seems oblivious to the meaning of what she is writing. Her meaning disintegrates in the next paragraph:

> In essence, I guess I have some prejudices against the in-group at Crestmont [the new church]. In my eyes they don't live up to the same level as my previous in-group. My attitude toward the pastor is not going to conform to the attitude of the group. I only "share" their feelings toward the leader. (Durst 1999, 156)

She is attempting to conclude while throwing in the last concept to which she remembers she was supposed to refer (Allport's central subject was the nature of prejudice), but her reference only vaguely resembles the kind of prejudice Allport was analyzing. Her "guess" reveals the nature of her looping reference to the concept of prejudice. She tries to give a reason for her prejudice with imprecise language ("live up to the same level") and then switches to the subject of her "attitude toward the pastor." She is attempting to use what she thinks might be school-level language ("conform to the attitude of the group") because that is how she interpreted the abstract language of writers like Skolnick and Allport. Almost as if conscious of her awkward attempt to speak in others' voices, she puts scare quotes around her next borrowing, "share." This appropriation of the word in fact throws her meaning off, for "share" semantically contradicts her statement that her attitude does not conform to the group's.

Louise tries to wrap up by formally reintroducing Allport into her essay. "Allport states," she writes, "no individual would mirror his group's attitude unless he had a personal need, or personal habit, that leads him to do so." Having misused several of Allport's words, she here misappropriates his claim so that it will fit her idiosyncratic interpretation of human behavior. In the particular passage from which this statement was lifted, Allport's claim is embedded within the larger claim that most of our behavior (and prejudices) are a consequence of our need to fit in (groupthink—not good). With some disruption in meaning, Louise shifts to an assertion that her needs are not that of the "in-group," asking the reader to travel back to the beginning of the previous paragraph where she mentions the "in-group at Crestmont." Louise's logic, which the reader must supply, is that Allport says we become members of an in-group when our needs are their needs; because she doesn't have their needs, she's not a member of the Crestmont in-group (Durst 1999, 156).

She wraps up the essay with writing that is part exploratory (she tries to clarify for herself why she's sticking around Crestmont) and part academic as she struggles to tie her exploration back to Allport's essay, using his words but bypassing his subject. "Just because I feel a part of the out-group," she concludes, "I'm not going to leave the church, it has a lot of positive factors [her attempt at academic lexicon] that I like and haven't been able to find anywhere else. If my needs ever do change I feel certain that I would be welcome into the in-group" (Durst 1999, 156).

For me, the most notable characteristic of this essay is the progressive disintegration as Louise moves from an interesting beginning to a broken attempt at mimicking academic writing. I see three layers of problems in the last half of her essay—the increasingly frequent problems with conventions and stylistics; problems with organization, coherence, and meaning; and psychological problems with the theme, assignment, and politics of the class. The first two layers are the tip of the iceberg; the last layer lies underneath. The submerged problem in part explains how the critical literacy agenda blocks Louise's development as a writer and consequently—at least to some extent—her progress toward her instrumental goal of becoming a nurse.

Durst (1999) claims that Louise's problem stems from her resistance to Stanforth's political agenda. Durst submits a secondary thesis that Louise and students like her have a serious problem with reading and writing in the genre of academic argument (130). I have been arguing along the lines of Durst's primary thesis, but I take exception to the secondary one. I think that the first is the cause of the illusion of the second. In a politicized classroom, teachers may see the second without understanding how it is rooted in the first.

I am suggesting that if we seriously hope to help students with their writing, we should revise our insistence on challenging their thinking. The critical thinking agenda seems to blur our focus on problems with style, grammar, and conventions, problems that might be exacerbated by our attempts to challenge their home values. Rarely in academic articles and books addressing the problem of student resistance to the critical literacy agenda do we see the kind of analysis I have attempted with Louise's essay. Our blurred focus may in fact account for a good deal of the legendary student resistance, because students come into our classrooms looking for help with their writing. Working-class students, in particular, do not expect and may not appreciate attempts to get them to rethink their religious, social, and political convictions. The consequence is allodoxia: students and teachers misreading each others' purposes, texts, and social locations. If one misreads a text, one misreads the problem and, as I am suggesting Durst did, puts the blame elsewhere.

8
THE PROFESSORS

The reader who complained about my dismissal of the importance of argument in the working-class ethos also thought I had stacked the deck in my critique of politicized writing instruction by focusing on graduate students and community college teachers instead of rhetoric and composition professors, claiming that members of the professoriate are better versed in the literature and consequently less likely to blunder than teachers like Cale, Stanforth, and Hendrix. Although I partially ascribe to my reader's logic, I don't entirely accept the positioned implication that more theory leads to better practice.

Nevertheless, in this chapter, I will focus on professorial critical teachers. Among them, Bill Thelin's (2005) account, "Understanding Problems in Critical Classrooms," is notable because he violates the naturalized narrative of failure, enlightenment, and triumph, a format for a genre that usually includes accounts of how others have gone wrong. Thelin leaves his analysis of his classroom failure mostly bare, grounding his article in the theory of failure on which he and John Paul Tassoni (2000) based their collection of "blunders."

Thelin (2005) narrates in explicit detail the degree to which his reading of liberatory pedagogy, as popularized by Ira Shor (1992, 1996), led to a classroom disaster. Thelin has taken seriously the methods Freire and Shor described, most of which fall under the rubric of democratizing education: co-developing generative themes, curriculum, contracts, and teacher-student responsibilities. Committed to a strategy of redistributing authority, Thelin holds up his own classroom practice to a critical examination in order to investigate with readers the missteps in his process. Thelin argues that we need to examine these missteps, interpreting them as a failure in strategy, not purpose. The smart mechanic doesn't junk the car simply because rebuilding the carburetor didn't solve the problem of acceleration hesitation.

With some differences, the breakdown in Thelin's (2005) class follows a model for failed critical teaching: too much time spent on co-development, student perception of teacher incompetence, missed classes (fewer than two-thirds of the class turned up on the second day), missed assignments, missed conference appointments, incomplete drafts, dysfunctional peer response sessions, student factionalism, student protest to upper administration, mid-course attempts at correction, and a hastily constructed assignment at the end of the semester directing students (the ones who were still there) to explain in five hundred words or more what had gone wrong with the class.

Thelin's desperate essay assignment reads like a cultural trope—an attempt at reconciliation on the brink of a failed marriage. Seitz (2004) describes a similar attempt to rescue a failed ESL class in which he "intended for students to mine their cultural experiences and anthropological reflections to develop academic habits of defamiliarizing the familiar" (19). After reading his students' essays, Seitz realized the students were, as he puts it, "dead tired of writing about their cultural selves" (19). After a what's-wrong-with-this-class? discussion, Seitz tried to revive the class by throwing his plans out the window and co-developing a new curriculum and themes to investigate. As a response to their failed class, Hendrix and Jacobson (2000) tried the same trick of having students co-develop a writing task and coming up short with a movie critique and the threadbare "24 hours to live," writing assignments that are about as lame as they come, indicating the students were only going through the motions, holding their breaths until the semester was over.

Once a teacher begins to flounder, students sense the teacher's lack of control, at which point there is little one can do to recapture their confidence in a teacher who doesn't seem to know what he or she is doing. Shor (1996) managed to pull his democratizing move from the brink with an on-the-spot invention of the now-famous "after-class" sessions when one of his students proposed that attendance should be voluntary (92-116), but few of us weather that moment when we have to get advice from students on how to fix our class. I remember my own anti-democratizing moment as a beginning high school teacher when three cowboys in twenty gallon hats got up and walked out after I announced this was their class, not mine.

Conscious of the academic injunction against unmitigated confession, Thelin (2005) attempts to ameliorate the depths of his "blunder" by contrasting it with a similar afternoon class that worked, the students co-developing the course and all happily passing the department portfolio assessment eleven out of the twenty students in his morning class failed. He also lays part of the blame for his failure on other teachers given to lock-step, process teaching, teacher-centered pedagogy, variations of the five-paragraph essay, and traditional grading systems. Although both Durst (2006) and I were less than convinced by Thelin's attempt to throw the blame elsewhere, I think Thelin located an important source of his failure when he says he neglected to investigate and honor the students' worlds adequately, the fundamental requirement of Freirean pedagogy. The frequency with which critical teachers miss Freire's ([1970] 1995) distinction between teachers who work from the students' worlds and teachers who impose their worlds on students is both instructive and curious (59-61). They also give liberatory pedagogy a bad name, leaving room for critics like Miller (1998) to create straw people who are only too easy to burn.

HUBRIS OR HUMILITY

Freire never shrank from acknowledging his expertise in his field, but he did not assume he had the right to lead his students to his sociopolitical perspective. Concomitantly, I am arguing with Fish (2008) that writing teachers should not deny their expertise about writing, but they should not generalize that knowledge, claiming that because they know more about A than students, they therefore know more about B through Z. If teachers limit their knowledge claims to their areas of expertise, they may focus on that area in their instruction; if they are beguiled into assuming preferred knowledge in B through Z, they may overshadow writing instruction by focusing on transformation, urging the students to assume socialist, libertarian, or conservative positions, depending on the teacher's perspective. In spite of their claims of ideological objectivity, teachers who move outside the area of their expertise are engaging in pedagogical hubris, the opposite of which is pedagogical humility, an announcement of the limits of one's own expertise.

Amy Lee (2000) and Tony Scott (2009), both skilled critical teachers, oscillate between hubris and humility. The organization of Lee's and Scott's books, are structural representations of this tension. In spite of Lee's observation that most articles or books by critical teachers are high on theory and low on practice (101), she spends the first six out of eight chapters focusing on theoretical issues, genuflecting to Foucault and replaying postmodern tropes about the personal being political, writers being written, defamiliarizing the familiar, and so on. Scott (2009) likewise overloads us with 130 pages of theory before he gets to 60 pages of sound pedagogical practice.

Lee (2000) has rooted her transformative pedagogy in the liberal arts tradition that imagines nurturing writers as the purpose of required writing courses. Teachers in this tradition feel the call to help others learn through writing who they are and where they want to go. Lee politicizes this tradition, interpreting writing as a fundamental mode of personal transformation within the social sphere. She writes in her introduction that she had originally set out "to write an argument about how writing and teaching writing can change the world" (1). Although she claims she learned to attenuate her utopic intent, the rest of her book suggests she didn't go far beyond it. She is clearly invested in getting her students to see the world as she does—and those who don't are not going to do well in her class, no matter how much she claims like other critical teachers (Graff 2001; France 1993; Lazere 1992; Strickland 1990) that she was creating a "safe" space where "differences, between students or between students and teacher, could be taken up and discussed, need not be repressed or glossed over for fear of abuse of power" (175).

Although on a different road, Scott (2009) is also driving hard toward social liberation. He more insistently grounds his pedagogy in a Marxist analysis of the effects of material conditions on social identity, hoping to sensitize students to the relationships between work situations and identity formation. Scott brings to the table a welcome refocus, inaugurated in 1986 at the Wyoming Conference, on the link between labor practices and pedagogy, an elephant generally kept in the closet for a host of disconcerting reasons associated with English studies and social class conflict (Slevin 1987; Sledd 1991; Bousquet 2002; Peckham 2009). Scott claims that we can't adequately address what we should be doing in our required writing programs without understanding the social, political, and economic frames within which we teach writing and the degree to which we as professors are reifying the social hierarchy we pretend to dismantle (see Peckham 1999). Scott focuses on the inherent contradiction in required writing programs—professors professing egalitarian social policies while benefitting from exploited labor, a contradiction Scott claims that cannot help but misshape writing instruction.

Scott (2009) and Lee (2000) follow the academic tradition of tearing down one's predecessors to make room for one's Promethean vision.

The protocol seems to demand simplifying one's predecessors' stances, which Lee and Scott do liberally. Lee reproduces the worn-out oversimplification of Hairston's (1992) protest against the politicization of writing instruction, pretending there are actually some teachers who believe that teaching can be apolitical (34). Following Faigley's (1992) declaration of the death of process pedagogy, both Lee and Scott critique process pedagogy as the "new formalism" (see Harris 1997), as if those of us who have been helping students understand that teaching writing is more than giving an assignment and then correcting the student's essay ever imagined writing as other than recursive and messy.

In keeping with his focus on the political economic, Scott (2009) aligns himself with Marc Bousquet's (2002) critique of writing program administrators as bureaucratic toadies. This critique depends on a false opposition between academia and business. Thus, academics like myself who are in academic/managerial positions are located as accommodationists whose message to our students is to scope out the dominant discourses and slip into them. The link between accommodationalism and genre-focused pedagogy is clear: teaching students how to write within genres is like teaching them to accommodate themselves to existing social formations (Peckham 1997). In the liberal arts tradition of Newkirk (1989) and Spellmeyer (1993), Lee and Scott encourage writing that pushes against formal limits, creating its own forms and within new forms, new meanings.

Having undercut a pseudo opposition, Lee (2000) and Scott (2009) describe their pedagogical goals in language that would have made Hairston shudder. Lee describes her need to "make visible the conditions by which knowledge is constructed" (36), "defamiliarize the production of meaning by presenting it as a construction, highly influenced by language, in which students can actively intervene,"(46), "contextualizing the possibilities and projects of critical pedagogy within a writing class, negotiating ways to engage students in an examination of the political implications and consequences of their texts, their positions as readers, their investments in being 'authors,' and the relations of a given classroom"(88). Scott also throws grace and clarity to the wind by framing his purpose within Herndl and Bauer's call for teachers to "'come into being' in politically creative and dynamic ways." Herndl and Bauer explain that "[w]hen those who had been excluded from the traditional norms of the universal usurp that position and speak as enfranchised subjects, the performative contradiction exposes the exclusionary nature of the conventional norm of universality and broadens the definition, creating a new space and subject position for the previously excluded" (Scott, 142). Scott also aligns himself with Min-Zhan Lu's "interventionist" pedagogy. Lu explains that "[t]o intervene with the order of Fast Capitalism, it is the responsibility of Composition to work with the belief that English is enlived—enlightened—by the work of users intent on using it to limn the actual, imagined, and possible lives of all its speakers, readers, and writers, the work of users intent on using English to describe and, thus, control those circumstances of their life designed by all systems and relations of injustice to submerge them" (Scott, 141).

I suspect that rhetoric like this has gained traction in our field because of the social class logic that privileges theory over practice (North 1987), the obscure over the plain-spoken (Bourdieu 1984); the field in its historical quest for legitimacy responds more toward the rhetoric of high theory than plain teaching. This rhetoric has also beguiled Lee (2000) and Scott (2009) into believing they are correcting student writing by correcting their thinking.

Lee (2000), for example, spends nine pages reflecting on the problem she had with a student's homophobic essay and her inability to lead the student "aright" (174). The prototypically white male student, Jay,[1] had written an essay arguing that homosexuals make bad parents. Lee contrasts Jay's essay with that of another student who wrote about her decision to let her mother know she, the student, was gay. Lee characterizes the gay student's essay as a resistant (and therefore heroic) discourse against the dominant discourse, which Lee claims the homophobic student represents. When evaluating Jay's essay, Lee claims she was able to turn her strong objections to his thesis aside and evaluate it on the basis of

> how his essay failed to make sense, how its logic broke down, how its examples and conclusions were not clearly related to one another. Because his conclusions relied on what *he already knew* to be 'true,' what he already knew to make sense, his essay was not making explicit this knowledge, but building out of it. So we took up what he knew to be true, and how it conflicted with what I know to be true. (176-177)

These justifications of a negative evaluation are an academic trope, the kinds of things teachers are prone to say about any student's essay that

1. Her pseudonym for "Jay" is revealing.

offends them, impelling them to critique the student's logic in a way they wouldn't critique an essay with a thesis with which the teachers wholeheartedly agreed: e.g., that homosexual parents can be as adequate as heterosexual parents. If teachers want to assign the poor grade on the basis of the ideas, they find a way of conflating the social logic with writing skills; thus, they critique the position under the guise of critiquing the writing, the corollary of changing the world through the agency of a writing class.

I do not doubt that Jay may have succumbed to the easy logic that too often provides a rationale for marginalizing homosexuality. Lee (2000), for instance, says that one of Jay's reasons supporting his claim was that when homosexuals openly kiss in public, they would be embarrassing their children, leading to, I assume, serious conflicts in the children between what they thought they should feel and what they did feel about their parents. In fact, Jay's essay sounds like one of the traditional sophomoric essays that we get from students when we ask them to take a for-or-against position on a controversial issue without having seriously researched the issue and the subtleties underwriting a range of conflicting positions.

In Lee's (2000) description of her response to Jay's essay, it seems as if her position ("what I know to be true") *is* true and Jay's job is to fit his beliefs into hers—which the discussion Lee describes makes clear he failed to do.

> In talking with this student, our discussion was heated on both sides; it was also long and frustrating. I did not leave feeling, 'Ah! I have won, changed his mind and led him aright.' I left wondering if it was even worth it to have engaged. And I mostly felt uncertain about whether or not I wanted a job in which, when confronted with raw anger and the abasement of human life, I had to contend with whether or not it was 'okay' to respond, to voice my objections. (174-175)

Lee's (2000) frustration reveals the degree to which she was no longer talking about Jay's ability to write, the announced subject of the course. Jay is clearly being forced into changing his belief system, which he refused to do. Consequently, Lee says she found the essay "unethical and unacceptable" (172), an evaluation, I suspect, that means F—or D at best. Jay's roommate had been right about what would happen if he handed it in: "Your teacher'll fail you," he had said (Lee, 173).

What's more noticeable than Lee's (2000) self-justifying tone in these nine pages is, as I have said about other critical teachers, that her description of the situation does not address problems with Jay's writing, although there are feints in that direction. One could argue (as Lee does) that she is "correcting" his writing by asking him "to interrogate or demystify his own thought processes or evaluations" by unpacking "the ideology that served to mask his real position, offering him the protection of a preexisting justification" (170-171). But her response would have been more useful if she had complimented Jay for recognizing that when he was making claims that had to be supported with reasons, he tried to supply them—i.e., one reason that homosexuals have difficulty with parenting is that when they kiss openly in public, they embarrass their children. She might then have encouraged him to imagine readers who might not accept his unspoken premise that children with homosexual parents would or should be embarrassed by an open display of affection when children with heterosexual parents would not. Lee might have even introduced him to the notion of an enthymeme here and requirements for readers and writers to share assumptions if an enthymeme is to work.

Scott (2009) gives a less problematic reading of one of his students' essays; nevertheless, his overall response raises the same question about the degree to which he allows a critique of a student's sociopolitical stance to infect his evaluation of her writing. Scott is acutely aware of the tendency of critical teachers to supplant writing instruction with discussions of values, but the vast bulk of his analysis of what his students learn centers on social values and the evolution from individualistic to systemic reasoning. He writes, for example, about Sophia, an African-American, first-generation student who seems on the verge of moving from accepting sexism in her specific work-situation to a systemic critique of gender discrimination. He describes her stance toward workplace policies in her first essay as interesting because she is able to identify the manager's blatant sexist policies without judging them. Sophia works in a restaurant in which the manager is male and the waitresses female. She says the manager freely admits that "sex sells"; consequently, he tends to hire attractive waitresses who know how to smile and flirt with their customers. Although Sophia makes some moves in her essay to suggest a critique, Scott notes that she does not seem unduly bothered about the role she plays in this cultural performance. Scott, however, is bothered, presumably about sexist social practices, but also about

Sophia's nonchalance and willingness to play her part, that is, to accept her instrumentalist, working-class function.

Scott (2009) narrates Sophia's partial evolution in a sequence of assignments from an individualistic to a systemic perspective. Scott attributes this evolution, rightly, I think, to a shift in genre from the auto-ethnography focusing on the experience of the self to the multi-media research essay leading to generalizations based on multiple experiences. Sophia's final paper, a multimodal essay addressing the false promise of advanced education, unnerves Scott because she concludes with a piece of advice that, although trite, seems to me close to the truth: "Networking is often the key to success, and if you master that, along with your college education, then gaining a position within a reputable company should be no problem; just make sure you have a last resort" (176).

Scott's (2009) response is as interesting as Sophia's response to her research. He says, "I nevertheless find where she ends up unsatisfying. The problem with this essay is that she wraps things up in a way that isn't true to the problems that it has raised" (178). "The truth" depends on how one sees the problem. Scott wants to see the problem from a social justice point of view, one that rejects networking as an ethical practice. As a working-class academic, Scott may be calling on his home Discourse in which sucking up marks "ear 'oles," one of which he and I have become. Sophia, however, on the other end of the social trajectory, might simply be trying to understand how the in-place social protocols work. She has in fact perceived that the promise of higher education is bogus[2]—that the real game is elsewhere, in your connections. If I would critique her thought, I would say she hasn't yet realized the full truth of the glimpse she has had, the degree to which she is almost inescapably handicapped in a race that privileges white, male middle- and upper-class students.

In spite of Scott's (2009) unease, I think Sophia is being the realist, just as she was when she accommodated herself to the sexism in her restaurant, the social structures within which she was able to turn her attractiveness and pleasant, flirting personality to her advantage. Scott resisted this sexist protocol, but both the adherence and resistance to sexist play are cultural productions. One could easily substitute physical prowess, intellectual acuity, or writing ability for the standard

2. Compare to Bourdieu's (1984) explanation of hystereisis (142).

of preference. From an attractive woman's point of view, why should culture give preference to writing ability? From a certain perspective, Sophia seems more in touch with the game than Scott, and game it she does, trying to give Scott what he wants just as she tried to accommodate her boss, Kevin. One has to remember, Sophia was just trying to get a good job, an instrumentalist motivation middle-class academics might not be able to understand. Scott's suspicion of his own problematic might be the consequence of a ghostly presence, memories of what he did to escape his working class. His preferences, he insightfully says, "risks having Sophia do school with me just as she does the happy worker at Shoney's" (178).

GOOD PRACTICE

Freire insists, that theory without action is "idle chatter, . . . *verbalism,* . . . an alienated and alienating 'blah'" ([1970] 1995, 68). More politely, I am suggesting that in the case of Lee (2000) and Scott (2009) that a surfeit of theory gets in the way of their good teaching. Scott in particular has designed a coherent sequence of assignments that build on each other to examine the theme of work, moving, as both Moffett (1968) and Freire recommend, from the inside to the outside, from the particular to the general, the existential to the sytemic. In Scott's class, students first write about and reflect on their lives as workers. Then they move outward to research information about the kind of work they do, record and analyze the discourses in their work situation, and finally embed this primary research in secondary research about work in and out of school. The students learn through their investigations more about the function of work and the hierarchy of kinds of work in our culture. Through learning more about "structuring structures" (Bourdieu 1984, 53), the students would inescapably learn more about the self, which is always the self-in-culture, or more explicitly, the self as a moving target in an array of conflicting and partially coinciding cultures.

Through this sequence of assignments featuring research and the incorporation of other voices into their own and allowing their voices to be reshaped by that incorporation, students are learning primary academic skills that they will be required to demonstrate in many of their other undergraduate classes with sustained writing assignments. In addition, Scott's sequence invites student investment because the students begin from an important part of their own worlds, particularly those students who spend a significant portion of their time working. In contrast to the subjects most critical teachers have their students investigate, this sequence may privilege working-class students, who have far more personal investment in the working world than students who come from the middle-middle or upper-middle classes, some of whom may never have worked for wages.

Lee's (2000) assignments are more problematic than Scott's (2009). They invite students to fight each other over issues of racial and gender identity, generating subjectivity wars over whose identity counts most (George and Shoos 1992, 203). Lee's male white students seemed to have picked up on the conflict right away, complaining about her first assignment in which she asked her students to write

> about assumptions they commonly encounter about their own identity, what groups they are typically assigned to, what characteristics are attributed to them as a result of that grouping, how these assumptions vary or remain the same according to context, who was most likely to buy into and act on these stereotypes, how these assumptions function not only to affect them individually but also to produce broader social implications, and how and why (and whether) individuals respond when they encounter them.

Lee (2000) was particularly critical of her four white students who said the topic "privileged the experience of those who were conscious of experiencing stereotypes" (111). Reflecting the popularity of "whiteness" theory in our field, Lee is critical of these students' inabilities to recognize their own relatively invisible markers of privilege, but I think the white students had a point—students who have been more clearly marked off from the dominant Discourse will not only be more invested in the topic but also be more able to recall poignant instances of the negative effects of stereotyping. I am not overly concerned about who is or isn't privileged by such a topic but rather by the divisiveness it invites, the kind of wrangling over identity politics that has eaten into social justice movements, distracting from the more important goal of working together to ameliorate structures of oppression. Lee's sequence of assignments designed to get students to revise their identity kits (143) clearly invites this kind of wrangling, leading to the angry confrontations in her class and between herself and her white male students and to her moments of wondering whether she had chosen the wrong profession (175).

I find the raw nerves exposed by Lee's (2000) pedagogy particularly troublesome, because she is otherwise a strong writing teacher. Her students, including, perhaps, the four or five white students with whom she seemed to have a running series of conflicts, learned something significant about writing, including strategies that would help them cope with their other undergraduate classes and take their places in the post-baccalaureate world—even those who gravitate toward Wall Street. Lee's students learned how to respond usefully to one another's essays, how to anticipate readers' responses, how to revise locally and globally, and how to adjust register and tone. They learned about the sociology of grammar, the relationship between power and language. They learned how to rely on each other for advice and how to respond to that advice. They had the affective experiences of learning through writing, of writing self-invested essays, and of having had their writing taken seriously, even if contested, by other readers.

I particularly appreciated Lee's (2000) focus on student texts rather than the works of professionals, a strategy Moffett (1968) recommended as an alternative to the corporately profitable modeling pedagogy depending on readers (208-210). Like Moffett and Elbow (1993), Lee minimizes the function of grades, infantilizing crutches directing student attention away from teachers' and other readers' discursive responses (Peckham 1993). In contrast to the common practice of teacher intervention in the writing process, Lee, like Thelin (2005) and Scott (2009), refuses to comment on her students' drafts, throwing the responsibility of being strong readers onto the students, a practice that not only makes for happier teachers but also encourages students to make writerly decisions as they weigh readers' different suggestions. I apologize for trying to rewrite Lee's book, but I would have liked to have read less about her theory and more about her teaching strategies and discussions of how they helped or failed to help her students become better writers. Maybe by becoming better writers, they might become better thinkers, and maybe by becoming better thinkers, they might become more socially conscious, but these relationships are far from a given, if Ezra Pound is any example.

Touchy Subjects

Our field has been in part defined by the tensions within the rhetorical situations we construct for student writers. The subject is a central feature of those situations. The tension within the subject is between student involvement and our responsibilities to teach particular writing skills, some of which seem to require the construction of rhetorical situations particularly uninviting to student writers. Richard Whately ([1846] 1963) was perhaps the first rhetorician to appeal to the student writer's interest when he constructed in 1828 "My Summer Vacation" as an antidote to the painful essays he had received from the more traditional topics like "Virtue, "Citizenship," or "Disobedience."

Brodkey's (1994) solution to this contradiction was to link the development of sociopolitical awareness to writing instruction. The issues of choice were racism, reverse racisms, and sexism as seen through the lens of affirmative action, the derivatives of which have dominated critical pedagogy and the conflict over the degree to which we should politicize writing instruction in required writing classes.

The benefit of this move has been to enliven the classroom. After facing class after class of disengaged students, nothing is more alluring to the frustrated writing teacher than to lob an issue like racial preferences into a classroom and watch it explode. Teachers can perhaps imagine themselves as effective because they see the students becoming involved in the arguments rather than staring out the windows while the teacher explains cohesion. I know how exciting it is to have students passionately engaged in a discussion about an issue they will investigate and write about, but I also know how easily these topics veer off-track when disputants are arguing about something more than the issue at hand, almost as if they are arguing about social spaces and one's right to be there (see, for example, Milanes 1991). By pursuing touchy subjects that ask students to "re-read America" through race, gender, and sexual preference lenses, teachers like Stanforth, Cale, Hendrix, Jacobson, and Lee have risked alienating groups of students and factionalizing the classroom.

High on the list of touchy subjects is religious fundamentalism. I acknowledge my tendency to raise my noetic eyebrow whenever students write about their unshakeable faith in Jesus as the literal son of a male God, or about Heaven and what it takes to get there. I am inclined to mention other candidates for Jesus' parentage and the difficulties with virgin births. This, after all, is higher education, a venue in which students need to think critically about creaking mythologies that have outlasted their times.

I am ending this chapter with a discussion of religious fundamentalism because of an article that Shari Stenberg (2006) wrote in defense of students whose religious beliefs tend to be dismissed by critical teachers. Stenberg, ironically, is the graduate student Lee (2000) describes as overly focused on a "radical feminist pedagogy" to the point of excluding writing as practice (98). Lee's mentoring and later collaborative work with Stenberg very likely helped Stenberg to refocus on the student as subject rather than object, a Freirean priority teachers with a transformative bent tend to get backwards. Stenberg takes seriously the dialogic basis of liberatory pedagogy, "sharing in knowledge (rather than denying the validity of one another's knowledge)" (280), no matter who the student is or what her beliefs.

Stenberg (2006) frames her discussion within Chris Anderson's partially fictive narrative of a teaching assistant who comes into his office to complain about a student writer who had used the occasion of her essay to announce her rebirth, concluding her essay with the following passage:

> Christ died on the Cross for my sins [Cathy concludes]. There is no way that I can repay Him for that, but I will try. I shall try to live my life fully for the Lord, and do His will. Hopefully, in doing this, I will also lead others to him. I know that this would make Him happy, because He loves every one of us and wants us to love Him and let Him come into our hearts. (Anderson 1989) [3]

Stenberg argues that from their academic, postmodernist subject position, teachers might misread essays in which students make a series of declarations like this (see also Seitz 2004, 165). As an example, she describes her initial response to a student's essay that disappointed her because it presented a different narrative than the one she had been acculturated as a critical theorist to expect. Stenberg notes that the beginning of Mary's essay made her anticipate the "enlightenment" model: I-thought-that; I-had-an-experience; having-seen-the-light, I-now-think-this (i.e., the Damascus trope). As Stenberg makes clear, we have as academics internalized this narrative so deeply that we have naturalized it to the point of being discomforted when a writer violates it. In response to the literacy narrative assignment, Mary had begun her essay by describing her Catholic upbringing, a seamless fabric of her family, church, and school. Mary then described how her previously unquestioned assumptions were challenged in college by her professors and her peers. It was at this point that Stenberg expected the enlightenment turn. Instead, she got the reaffirmation narrative. Stenberg writes that "when it became what I didn't expect, or what I didn't want it to be, I found myself wanting to dismiss this conclusion as too easy or too simple" (283)—or as Scott (2009) might

3. Anderson (1989) says he is "fictionalizing a little here, but only a little" in his discussion of the religious student's essay and the annoyed teaching assistant.

have said, her essay "wraps things up in a way that isn't true to the problems that it has raised" (178).

I suggested above that Lee's (2000) hypercritical response to her homophobic student's essay may in part have been a consequence of a similar disruption of expectations, leading Lee to work overtime to discount the student's logic and examples. Stenberg (2006) points out that if read from Mary's frame of reference, her essay would have made perfect sense—that is, by having to think more deeply about her faith when it was challenged, she reaffirmed it. This makes perfect sense to me if I imagine the temptations (challenges) of Christ as the narrative frame.

Stenberg questions her right to "steer [Mary's] paper in the direction of my values" (284), which is precisely what Lee was trying to do to her student in the long and "heated discussion" that went nowhere. Seitz (2004) describes a similar attempt to steer a religious student's essay in the direction his immersion in cultural studies had taught him to steer it. Upon reflection, Seitz wondered whether his desire to get religious students like Kisha to complicate their assumptions was at least in part a self-privileging play "to maintain my field's critical 'standards' and so my particularly professional authority" (17). I would throw the weight of Seitz's question toward the need to maintain professional authority. We are always telling our students to interrogate information or arguments by asking who benefits, a question we need to ask of our own classroom strategies and performances. By doing X, do we really have our students' best interests in mind—or are we engaging in an action only as a cover for another point we're trying to make or a power relationship we think is important to maintain? So what do we get, Seitz suggests, out of asking working-class, religiously-oriented African-Americans to deconstruct their stories about faith, discipline, community, and hope—the values that sustain them in their struggles to stay off the street?

A few years ago, I was sitting next to a very pleasant-looking young woman on my flight home from a convention in Chicago. She said her name was Lynette. She had straight brunette hair, hazel eyes, unblemished skin, and an open look on her face—the kind of face that told you this young woman had never seen trouble. I mistook her for a high school student; she told me with no hint of having taken offense that she was a senior in college—some Baptist college in Oklahoma City. She and her family and other friends from their church were returning from Amsterdam, where they had spent their week of spring vacation on the streets urging strangers to invite Jesus into their lives.

Lynette was clear about her love for Jesus, which was obviously an outward manifestation of the love she had received from her parents and eleven siblings—spread in small groups throughout the plane. I interpreted her micro-love found within her family as regenerating itself through a macro-love for Jesus, the symbol for the intersecting worlds of spiritual and material reality. This young woman was planning to go to Japan where she wanted to teach English and spread her word. After a few years in Japan and other countries, she expected to return to the United States and go to a Baptist graduate school, where she would get a Masters in ESL. She had some vague ideas about going on for a doctorate, but she didn't know what that world was about. What she knew was that she wanted to teach—and have a family—and share her warm feeling about life.

In spite of her naiveté, I wish we had many more people like Lynette—I almost wish I were a bit more like her myself, for I know myself as layered in contradictions, still a bit even at my age like Joyce's fictional self described at the end of "Araby." At any rate, I thought about Lynette being in a class, as Stanforth wisely called it, engineered by the disassembly crew (Durst 1999, 169). No, I wouldn't want her there. There is no earthly reason to disassemble Lynette's beliefs, to expose her to the possibility that the story of Jesus is mostly story, to alert her to the functions of grand narratives, the illusion of identity, the secret engines of capitalism that run on the exploitation of marginalized social groups. If I had Lynette in my class, I would think it wrong of me to attempt to get her to see life though our kaleidoscope—now frequently disguised as a way of alerting our students to different ways of seeing. I wouldn't want to change her thinking. I would want to help her with her writing in a class as depoliticized, as de-ideologized as possible. She would go out with her own thinking and do good in the world. It would have been my pleasure to have helped her with her writing.

9
GOING WEST

I write with irony about instruction in required writing classes because I rarely teach them. What few courses writing program administrators in doctoral intensive universities teach are mostly graduate courses—or in my case, courses to teach new writing teachers how to teach writing. Given my subject of social class relationships in writing instruction, the irony doubles because I am a theorist and administrator writing about how elites maintain social privileges through distancing themselves from the real work in culture. In our subculture of rhetoric and composition, the real work is teaching English 101. I view this work from the distance of my office with wide windows.

When we read about the actual dramas in classrooms, the teachers are overwhelmingly non-tenure track instructors or graduate students. As a tenured professor with a PhD in rhetoric and composition, I can't address the conflicts in this situation without acknowledging myself as a privileged spectator. Like my critical reader, professors tend to assume that the more knowledge one has about a subject, the more one is qualified to teach it. Because we have had more leisure time than instructors or graduate students to study writing instruction, there is some presumption that—all other things being equal—we will be better writing teachers than those who have been submerged in the daily work of teaching, those who never get time to step back to research and reflect on what they are doing (see Seitz 2004, 14). The implication to social classes and living style seems painfully obvious.

Because of the work situation, instructors tend to be long on practice and short on theory; the GTAs are the reverse. Our work as WPAs is to blend theory and practice, to do our best to ensure effective instruction for our undergraduates in spite of this mix and the exploitive labor mechanism underwriting it. My primary function as a WPA is to encourage pedagogies that will help students from all social groups negotiate the writing demands in their undergraduate classes, regardless of the use to which these students put their improved writing skills.

The general critique of this instrumentalist pedagogy is that as educators, we have a double function—to help students with their writing *and* prepare them to be critical citizens in a participatory democracy. The conundrum lies in this mix—or rather, in what the mix is. When the critical literacy project impinges on the writing skills project, how do we balance the imperatives and at what point do we backpedal on one to save the other—*if* the two are in conflict?

The continuing controversy over this question makes clear that the issue is far from settled. Richard Fulkerson's (2006) summary of current trends in rhetoric and composition stimulated a protest from several respondents who objected to Fulkerson's treatment of what he called the critical/cultural studies (CCS) approach. The response to Fulkerson was intemperate in one of our regular discussion groups that we hold for writing teachers at Louisiana State University. The dialogue was equally as heated in the exchange between William Thelin (2006) and Russel Durst (2006), following Thelin's 2005 article that I analyzed above. Responding in part to Thelin's implication that Durst's pedagogy of "reflective instrumentalism" was a variety of "status-quo teaching" (111), Durst fired back with a charge that Thelin was looking everywhere but at his own pedagogy in an attempt to discover why Thelin's class failed.

The Thelin/Durst exchange was perhaps a reminder not only of the enduring relevance of the issue but also of its counterproductive intensity, making people who would otherwise work together work against each other. Although Thelin and Durst disagree about the link between required writing programs and politicized instruction, they share similar sociopolitical and pedagogical orientations. Likewise, I share a sociopolitical philosophy with many of the teachers I have critiqued in my discussions of the social strand of critical thinking. We might be described as progressives, socialists, communitarians, neomarxists, or even liberals—if we can manage to overcome that label's link with the autonomous individual. We believe in working for the good of the community and we disapprove of hoarding. We do not approve of the capitalist ethos and its mindless celebration of competition. We know that the educational system tends to reproduce the social system within which it exists. We believe that we must challenge all institutional mechanisms that maintain systems of inequity and oppression, both inside and outside our educational culture. To challenge these mechanisms, one has to learn how to spot unsubstantiated claims and how to read underneath social practices that train us to accept as natural the structuring structures that are the fabric of our culture. With so much in common, we need to work together. We should listen carefully to each other and learn how to give ground to find common spaces.

I have described my position on the issue of politicized writing instruction, but I need to emphasize that I have never imagined myself as standing on solid ground. I change as I listen to my friends, and I change as the political climate changes. I began this book before the Bush administration invaded Iraq, encumbering this nation in a reckless military adventure that has by now cost well over 150,000 lives, over 730 billion of our tax dollars and still counting, our international reputation, one destroyed country, and countless disabled veterans and displaced Iraqis—all the consequence of our having twice put in office a president who turned everything he touched into sludge. My experience in watching and resisting the Bush administration has certainly made me reflect on our failure in educating critical citizens, people who will question claims and search for corroborating and conflicting accounts of the evidence supporting those claims. I have understood that as writing teachers, we need to teach students not only how to write in academic genres, but we also need to teach them how to evaluate other discourses in order to enter intelligently into any serious public discussion.

If it is possible to interweave critical thinking in the cognitive and social strands with the more focused instrumental objective, as I have described it, I am all for it. I might even edge out on the limb and say that writing instruction that furthers the goal of a more tolerant and egalitarian society without impeding writing instruction or disadvantaging students from any social group, I might consider it, but I worry about providing a justification for teachers who ascribe to a more conservative ideology also to use their classrooms as pulpits for their beliefs. I believe in social justice as Paulo Freire describes it, but I can imagine and respect a writing teacher who is an anarchist, a libertarian, a Keynesian, a Buddhist, a Christian fundamentalist, a doctrinaire Catholic. I don't like to think of our required writing classrooms as a place where people with these different persuasions try to get their students to see life as they do.

MAKING A DIFFERENCE

At the level of personal profit, I am deeply suspicious of the degree to which I and other academics have been socialized into caring whether our words make a difference, but at the level of community, it is important that we do. We need to speak with our words and actions to contribute to our culture, even though we find ourselves in disagreement with our friends and neighbors. Although I am hyperconscious about the tension between working for personal and community profit and the degree to which the former gets masked as the latter and thereby legitimated,[1] I have committed myself to this field in which we write to make a difference, to alter the outer reality into which we push our texts. If we can give students the sense of their right to speak and a belief in the power of their words, we can walk away from our classes thinking we have done good work.

I began writing this book several years ago in reaction to a statement in a required writing review committee meeting by one of my colleagues that we should make critical thinking the focus of our required writing course. Heads nodded, and I exercised administrative restraint. The flip side of our duty to speak is knowing when not to speak—which might be, as McPeck (1981) has described it, an act of critical thinking. The rhetorical context of that situation was that I was a freshly imported WPA chairing the review committee. As I said, heads were nodding.

But that comment—and my decision not to speak—led me to track down what people meant by critical thinking, which research confirmed a suspicion that I had in the mid-seventies when the term started to replace at the high school level a fascination with the five-paragraph essay. I thought of it then as five-paragraph thinking, and my research, perhaps predictably, didn't prove me far wrong.

I have described in this book what I learned about critical thinking. In sum, I have described critical thinking as a rhetorical move, positioning the critical-thinkers against the non-critical-thinkers within specific contexts disguised as a general condition. The "critical-thinkers" are always those who have gained (or who thought they have gained) a surplus in legitimated educational and cultural capital within those contexts, a short-hand way of saying those who were born into a way

1. See Swartz's interpretation of Bourdieu's thesis of symbolic violence: "The logic of self-interest underlying all practices particularly those in the cultural domain is misrecognized as the logic of 'disinterest.' Symbolic practices deflect attention from the interested character of practices and thereby contribute to their enactment as disinterested pursuits. This misperception legitimizes these practices and thereby contributes to the reproduction of the social order in which they are embedded. Activities and resources gain in symbolic power, or legitimacy, to the extent that they become separated from underlying material interests and hence go misrecognized as representing disinterested forms of activities and resources" (90).

of thinking while thinking they gained it—or to paraphrase what Jim Hightower said about George H. W. Bush, those who were born on third base and thought they had hit triples.

I have linked notions of critical thinking with other barriers that deter working-class students from succeeding in the upper levels of our educational system. I am mindful of the other social groups marginalized by parallel strategies—one could easily extrapolate from most of the excluding mechanisms I have described as social-class based correlate strategies that work against other dominated social groups. I am also wary of the implications of a dominant/dominated opposition—I know that social arrangements are not that simple, but we can't ignore the social reproductive mechanisms that created a space for a president like George W. Bush. Put less politically, we can't assume a significantly low percentage of working-class students get PhDs because they are not as intelligent as children in the middle and upper classes.

I have been particularly concerned in my analysis with pedagogical approaches that promote social justice but ironically penalize working-class students. My commitment to these same values complicates my concern, but I know they are in conflict with many of the values of my home community and other working-class students. In my cynical moments, I have wondered whether the postmodernist theorizing that dominates our discipline's conversations is only one more social reproduction mechanism, valorizing theory and leading away from the pragmatics of empirical research on effective teaching strategies.

I was particularly disturbed by Lynn Bloom's 1996 article, "Freshman Composition as a Middle-Class Enterprise," not only because Bloom wrote it, but because *College English* published it and that few scholars challenged what to me was a startling dismissal of the working-class ethos in that essay. I can't overstate how much I respect Lynn Bloom, but she and I grew up in such different worlds leading her to see white where I saw black. In part *Going North* grew from this disturbance. How could she so completely see one thing and I another with both of us imagining ourselves as critical thinkers?

As a consequence of these intersecting concerns, I explored in this book the many ways in which critical literacy teachers may unintentionally marginalize working-class students—acknowledging the position within which my social class experience has framed me. Language codes are one of the more obvious mechanisms of social class reproduction. The notion of critical thinking, as it is theorized by middle-class writers, is more covert. I have consequently made these mechanisms the focus of the latter portion of this book. The most interesting features of the covert mechanisms are the ways in which they make use of donor pedagogues, teachers who unknowingly carry the disease to unsuspecting recipients. These are working-class academics who embrace pedagogies that exclude students who share their social class origins.

I have divided critical thinking into the cognitive and social strands—both of which have strongly influenced writing instruction, the former through positing argumentative writing as the paradigmatic academic discourse, the latter through linking critical thinking with social justice. Although masked as the prototype of academic discourse, I suspect that argumentative genres have gained a central role in required writing because of their reputation in a hierarchy of genres in alignment with social classes along personal/impersonal and concrete/abstract continua. I have explored the reasons argument is consequently in congruence more with middle-class than with working-class students' habitus. My purpose has not been to exclude argumentative genres from our required writing programs but rather to alert teachers to the possible dissonance between argument and working-class students and to focus more on claims as rhetorical moves within many kinds of genres rather than on argument as the prototypical academic essay.

As Richard Fulkerson (2005) has argued, the social strand of critical thinking has gained a central position in required writing instruction in our publications, if not in our classrooms. Although popular, the call to employ our required writing classrooms as a platform for social change has predictably met resistance from students and teachers who want to keep the focus in required writing classes on writing instruction. In interpreting some of the written accounts of student resistance to teachers who have embraced the transformative imperative, I have taken a Freirean perspective toward students' purposes: that is, that teachers should understand and honor students' purposes just as we would like to have them understand and honor ours. As I have been disturbed by the middle-class perspective argued by Bloom (1996), I have also been disturbed by critical teachers who frame their pedagogy within an assumption of student naiveté. This kind of teaching assumes a privileging opposition of teacher as knower against students as non-knower, a classic domesticating strategy.

Like so many people in rhetoric and composition, I began my career teaching literature, but my experience in the 1960s and an interest in social class relationships made me question the elitism involved in notions of "writers" and people (like me) who were teaching literature as a way of identifying with them. I began to think of writing as an act rather than an identity, which ineluctably led me as a high school teacher to the Bay Area Writing Project, where I met Jim Grey, Miles Myers, Keith Caldwell, Mary K. Healey, Fran Claggett, Rebecca Caplin, Bill Robinson, and a host of other wonderful writing teachers. My experience with BAWP in 1977 was my personal transformation, one I have never regretted. Except for the many times in the beginning and the intermittent times throughout the rest of my career when my classroom strategies imploded, I have never wanted to do anything else but teach "just" writing. I know that by helping my students with their writing, I have made a difference in some of their lives. This has been my West. I may have ended up in Louisiana, but I have been going West in my mind.

REFERENCES

ACT. 2003. English and writing. *Content validity evidence in support of ACTs educational achievement tests,* http://www.act.org/research/curricsurvey.html (accessed February 5, 2010).

AFL-CIO. 2009. 2009 Executive pay watch. *Corporate watch,* http://www.aflcio.org/corporatewatch/paywatch/ (accessed February 5, 2010).

Aldrich, Nelson. 1996. Old money: The mythology of American's upper class. In *Created equal: Reading and writing about class in America,* ed. Benjamin DeMott, 212-24. New York: Harper Collins. (Orig. pub. 1988.)

Allison, Dorothy. 1993. *Bastard out of Carolina.* New York: Plume/Penguin.

Allport, Gordon. 1992. Formation of in-groups. In *Rereading America.* 2d ed. ed. Gary Colombo, Robert. Cullen, and Barbara Lisle, 292-306. Boston: Bedford Books of St. Martin's Press. (Orig. pub. 1954).

Alspaugh, John. 1992. Socioeconomic measures and achievement: Urban vs. rural. *Rural Educator* 13 (3), 2-7.

Althusser, Louis. 1984. *Essays on ideology.* London: Verso.

Anderson, Chris. 1989. The description of an embarrassment: When students write about Religion. *ADE Bulletin* 94 (Winter), http://www.ade.org/bulletin/index.htm (accessed February 5, 2010).

Anderson, Virginia. 2000. Property rights: Exclusion as moral action in "the battle of Texas." *College English* 62 (4): 445-72.

Anyon, Jean. 1980. Social class and the hidden curriculum of work. *Journal of Education* 162 (2): 67-92.

Apple, Michael. 1982. *Education and power.* Boston, MA: Routledge & Kegan Paul Ltd.

Aristotle. 1954. *The rhetoric and the poetics of Aristotle* . Trans. W. Rhys Roberts and Ingram Bywater. New York: Modern Library.

Arnold, Matthew. 1913. The function of criticism at the present time. In *Selections from the prose works of Mathew Arnold.* Ed. William Savage Johnson. New York: Houghton Mifflin, http://www.gutenberg.org/etext/12628 (accessed July 16, 2008).

Aronowitz, Stanley. 1997. Between nationality and class. *Harvard Educational Review* 67 (2): 188-207.

———. 1999. *Work and the politics of identity.* Keynote address. Working Class Studies Conference. Youngstown, Ohio. June 9-12.

Bakhtin, Mikhail Mikhailovich. 1981. *The dialogic imagination.* Ed. Michael Holquist. Trans. Michael Holquist and Caryl Emerson. Austin, TX: University of Texas Press.

———. 1986. The problem of speech genres. In *Speech genres and other late essays,* ed. Michael Holquist and Caryl Emerson, trans. Vern W. McGee, 60-102. Austin, TX: University of Texas Press.

Baker, Sheridan. 1962. *The practical stylist.* New York: Crowell.

Bartholomae, David. 1985. Inventing the university. In *When a writer can't write,* ed. Mike Rose, 134-65. New York: The Guilford Press.

Bartholomae, David and Anthony Petrosky. 1993. Preface to Bartholomae and Petrosky, eds. 1993, v-xiii.

Bartholomae, David and Anthony Petrosky, eds. 1993. *Ways of reading,* 3rd ed. Boston: Bedford Books of St. Martin's Press.

Bartolome, Lilia. 1998. *The misteaching of academic discourses: The politics of language in the classroom.* Boulder, CO: Westview Press.

Bauer, Dale. 1990. The other 'f' word: Feminist in the classroom. *College English* 52 (4): 385-96.

Berger, Peter. and Thomas Luckmann. 1967. *The social construction of reality.* New York: Anchor Books.

Berlin, James. 1987. *Rhetoric and reality: Writing instruction in American colleges, 1900-1985.* Carbondale, IL: Southern Illinois University Press.

———. 1988. Rhetoric and ideology in the writing class. *College English* 50 (5): 477-94.

———. 1991a. Transcript. In Hurlbert and Blitz 1991, 137-138.

———. 1991b. Composition and cultural studies. In Hurlbert and Blitz 1991, 47-55.

———. 1994. Poststructuralism, cultural studies, and composition. In *Professing the new rhetorics,* ed. Theresa Enos and Stuart Brown, 446-60. Boston: Blair Press.

———. 2003. *Rhetorics, poetics, and cultures: Reconfiguring college English studies.* West Lafayette, IN: Parlor Press. (Orig. pub. 1996.)

Berlin, James, and Michael Vivion, eds. 1992. *Cultural studies in the English classroom.* Portsmouth NJ: Boynton/Cook.

———. 1992(a). In *Cultural Studies in the English Classroom,* ed. J. Berlin and M. Vivion . Portsmouth NJ: Boynton/Cook.

———. 1992(b). "Poststructuralism, Cultural Studies, and the Composition Classroom: Postmodern Theory in Practice." *Rhetoric Review* 11 (1): 16-33.

Bernstein, Basil. 1971. *Class, codes, and control I.* London: Routledge & Kegan Paul.

———. 1996. *Pedagogy, symbolic control and identity: Theory, research, critique.* Bristol, PA: Taylor & Francis.

Beyer, Barry. 1985. Critical thinking: What is it? *Social Education* 49 (5): 270-76.

Bird, Barbara, Doug Downs, and Elizabeth Wardle. Downs and Wardle redux. 2008. *College Composition and Communication* 60 (1): 165-81.

Bizzell, Patricia. 1994. 'Contact zones' and English studies. *College English* 56 (2): 163-69.

Black, Max. 1952. *Critical thinking: An introduction to logic and scientific thinking.* New York: Prentice Hall. (Orig work pub. 1946.)

Bloom, Lynn. 1996. Freshman composition as a middle-class enterprise. *College English* 58 (6): 654-75.

Bourdieu, Pierre. 1984. *Distinction: A social critique of the judgement of taste.* Trans. Richard Nice. Cambridge, MA: Harvard University Press.

———. 1991. *Language and symbolic power.* Ed. John Thompson. Trans. Gino Raymond and Matthew Adamson. Cambridge, MA: Harvard University Press.

Bourdieu, Pierre, and Jean-Claude Passeron. 1990. *Reproduction in education, society, and culture.* Trans. Richard Nice. Newbury Park, CA: Sage.

Bousquet, Marc. 2002. Composition as management science: Toward a university without a WPA. *JAC* 22 (3): 493-526.

Bowles, Howard, and Samuel Gintis. 1976. *Schooling in capitalist America.* New York: Basic Books.

Breen, Richard, and David Rottman. 1995. *Class stratification: A comparative analysis.* New York: Harvester Wheatsheaf.

Brice Heath, Shirley. 1983. *Ways with words: Language, life, and work in communities and classrooms.* New York: Cambridge University Press.

Britton, James, Tony Burgess, Nancy Martin, Alex McLeod, and Harold Rosen. 1978. *The development of writing abilities (11-18).* Urbana IL: National Council of Teachers of English.

Brodkey, Linda. 1994. Making a federal case out of difference: The politics of pedagogy, publicity, and postponement. In *Writing theory and critical theory*, ed. John Clifford and John Schilb, 236-61. New York: Modern Language Association.

———. 1994. Writing on the bias. *College English* 56 (5): 527-47.

Brown, Rexford. (1978). "What we know now and how we could know more about writing ability in America." *Journal of Basic Writing* 1 (4): 1-6.

Cale, Gary. 2001. *When resistance becomes reproduction: A critical action research study*. Paper presented at the *Pedagogy and Theatre of the Oppressed Conference,* Omaha, NE. March.

Christopher, Renny. 1998. *Longing fervently for revolution.* Niagara Falls, NY: Slipstream.

Clark, Burton. 1960. The cooling out function in higher education. *American Journal of Sociology* 65 (6): 569-76.

College Board. 2004. *A guide to the new SAT essay*. New York: College Entrance Examination Board.

Colombo, Gary, Robert Cullen, and Bonnie Lisle, eds. 1992. *Rereading America,* 2nd ed. Boston, MA: Bedford Books of St. Martin's Press.

Conference on College Composition and Communication. 1974. *Background statement: Students' right to their own language,* http://www.ncte.org/ccc/ex.html/ (accessed March 3, 2004).

Cooper, Charles, and Lee Odell. 1977. *Evaluating writing.* Urbana, IL: National Council of Teachers of English.

Corzo, Patricia. November 13, 2000. Excellent book for foreign people. Review of *Rereading America,* http://www.amazon.com (accessed October 3, 2008).

Cosby, Bill. 1968. To Russell, My Brother, Whom I Slept with. *To Russell, My Brother, Whom I Slept with.* Record. Warner Brothers.

Counterstatement . 1993. *College Composition and Communication* 44 (2): 248-57.

Daniels, Jim. 1985. *Places everyone.* Madison, WI: University of Wisconsin Press.

Davies, Brian. 1995 . Bernstein on classrooms. In *Discourse and reproduction: Essays in honor of Basil Bernstein,* ed. Paul Atkinson, Brian Davies and Sara Delamont, 137-157. Cresskill, NJ: Hampton Press.

Dewey, John. 1963. *Experience and education.* New York: Collier. (Orig. pub.1938.)

Dews, Barney, and Caroline Law, eds. 1995. *This fine place so far from home: Voices of academics from the working class.* Philadelphia, PA: Temple University Press.

Dickens, Charles. 1958. *Hard Times.* Toronto: Rinehart and Co. (Orig. pub. 1854.)

Downs, Douglas and Elizabeth Wardle. 2007. Teaching about writing, righting misconceptions: (Re)envisioning "First-year composition" as "Introduction to writing studies. *College Composition and Communication* 58 (4): 552-84.

Durkheim, Emile. 1915. *Elementary forms of the religious life.* London: Allen and Unwin.

Durst, Russel. 1999. *Collision course: Conflict, negotiation, and learning in college composition.* Urbana, IL: National Council of Teachers of English.

———. 2006. Interchanges: Can we be critical of critical pedagogy? *College Composition and Communication* 58 (1): 110-14.

Eggleston, Edward. 1913. *The Hoosier schoolmaster: A story of backwoods life in Indiana.* Rev. ed. New York: Grosset & Dunlap.

Elbow, Peter. 1993. Ranking, evaluating, and liking: Sorting out three forms of judgment. *College English* 55 (2): 187-206.

Ellsworth, Elizabeth. 1989. Why doesn't this feel empowering?: Working through the repressive myths of critical pedagogy. *Harvard Educational Review* 59 (3): 297-324.

Ennis, Richard. 1962. A concept of critical thinking. *Harvard Educational Review* 32 (1): 81-111.

Enos, Richard. 1976. The epistemology of Gorgias' rhetoric: A re-examination. *The Southern Speech Communication Journal,* 42 (1): 35-51.

Faigley, Lester. 1992. *Fragments of Rationality: Postmodernity and the subject of composition*. Pittsburgh: University of Pittsburgh Press.

Farber, Jerry. 1970. *The student as nigger*. New York: Pocket Books.

Faulkner, Carol. 1998. Truth and the working class in the working classroom. In Shepard, McMillan, and Tate 1998, 37-44.

Finn, Patrick. 1999. *Literacy with an attitude*. Albany, NY: State University of New York Press.

Fish, Stanley. March 29, 2002. Is everything political? *The Chronicle of Higher Education*, http://chronicle.com/jobs/news/2002/03/2002032901c.htm (accessed December 18, 2008).

———. 2008. *Save the world on your own time*. New York: Oxford University Press.

Fort, George. 1884. *A critical inquiry into the condition of the conventional builders and their relations to secular guilds in the middle ages*. New York: J. W. Bouton.

Fox, Catherine. 2002. The race to truth: Disarticulating critical thinking from whiteliness. *Pedaogy*. 2(2). http://muse.jhu.edu/journals/pedagogy/toc/ped2.2.html. (accessed June 14, 2010).

France, Alan. 1993. Assigning places: The function of introductory composition as a cultural discourse. *College English* 55 (6): 593-609.

Frey, Olivia. 1998. Stupid clown of the spirit's motive: Class bias in literary and composition studies. In *Coming to class*, ed. Alan Shepard, John. McMillan, and Gary Tate, 61-78. Portsmouth NY: Boynton/Cook.

Freire, Paulo. 1993. Introduction to *Paulo Freire: A critical encounter*, ed. Peter Leonard and Peter McLaren, ix-xii. London: Routledge.

———. 1995. *Pedagogy of the oppressed*, 20th-Anniversary Edition. New York: Continuum. (Orig. pub. 1970.)

Fulkerson, R. (1990). "Composition theory in the eighties: Axiological consensus and paradigmatic diversity." *College Composition and Communication* 41 (4): 409-29

Fulkerson, Richard. 2005. Composition at the turn of the twenty-first century. *College Composition and Communication* 56 (4): 654-87.

Garcia, Valeriano. 1973. *A Critical Inquiry into Argentine Economic History*. n.p.: n.p.

Garger, Stephen. 1995. Bronx syndrome. In Dews and Law 1995, 41-53.

Gebhardt, Richard. 1992. Editor's column. *College Composition and Communication* 43 (2): 295-96.

Gee, James Paul. 1998. Introduction to *The misteaching of academic discourses: The politics of language in the classroom* by Lilia Bartolome. Boulder CO: Westview Press.

———. 1996. *Social linguistics and literacies*. 2nd ed. Bristol PA: Taylor & Francis.

———. 2004. Learning language as a matter of learning social languages within discourses. In *Language learning and teacher education: A sociocultural approach*, ed. Margaret R. Hawkins, 13-31. Clevedon UK: Multilingual Matters.

George, Diana, and Shoos, Diana. 1992. Issues of subjectivity and resistance: Cultural studies in the composition classroom. In Berlin and Vivion, 200-10.

Giroux, Henry. 1991. *Border crossing: Cultural workers and the politics of education*. New York: Routledge.

———. 1993. *Living dangerously*. New York: Peter Lang.

Glaser, Edward. 1941. *An experiment in the development of critical thinking*. New York: Teachers College, Columbia University.

Glaser, Edward, and Goodwin Watson. 1980. *Watson-Glaser critical thinking appraisal*. San Antonio: The Psychological Corporation/Harcourt Brace Johanovich. (Orig. pub. 1941.)

Graff, Gerald. 1992. *Beyond the culture wars: How teaching the conflicts can revitalize American education*. New York: W.W. Norton.

———. 2001. Teaching politically without political correctness. *Democratic Culture* 6 (3): 3-8.

Gramsci, Antonio. 1971. *Selections from the prison notebooks of Antonio Gramsci.* Ed. and trans. Quintin Hoare and Geoffrey Nowell Smith. New York: International Publishers.

Hairston, Maxine. 1990. Comment and response. *College English* 52 (6): 694-96.

———. 1992. Diversity, ideology, and teaching writing. *College Composition and Communication* 43 (2): 179-93.

Hardin, Joe Marshall. 2001. *Opening spaces: Critical pedagogy and resistance theory in composition.* Albany: State University of New York Press.

Harris, Joseph. 1997. *A teaching subject: Composition since 1966.* Upper Saddle River NJ: Prentice Hall.

Hart, Betty, and Todd R. Risley. 1995. *Meaningful differences in the everyday experiences of young American children.* Baltimore: Paul H. Brookes Publishing Co.

Havelock, Eric. 1963. *Preface to Plato.* Cambridge: Harvard University Press.

Hendrix, Scott., Sarah Hoskinson, Erika Jacobson, and Saira Sufi. 2000. What happened in English 101? In Thelin and Tassoni 2000, 51-67.

Johnson Black, Laurel. 1995. Stupid rich bastards. In Dews and Law 1995, 13-25.

Katz, Michael. 1975. *Class, bureaucracy and schools: The Illusion of educational change in America.* New York: Praeger.

Kerbo, Howard. 1996. *Social stratification and inequality: Class conflict in historical and comparative perspective.* 3rd ed. New York: McGraw-Hill.

Kinneavy, James. 1969. *A theory of discourse.* New York: W.W. Norton and Company.

Knoblauch, C. H. 1991. Critical teaching and dominant culture. In Hurlbert and Blitz 1991, 12-21.

Kozol, Jonathan. 1991. *Savage inequalities.* New York: HarperCollins.

Kruse, Kim. 1996. The effects of a low socioeconomic environment on a student's academic achievement. Research Report. ERIC: ED402380.

Kuhn, Thomas. 1970. *The structure of scientific revolutions.* Chicago: University of Chicago Press.

Lakoff, George. 1987. *Women, fire, and dangerous things: What categories reveal about the mind.* Chicago: University of Chicago Press.

Lareau, Annette. 2003. *Unequal childhoods: Class, race, and family life.* Berkeley, CA: University of California Press.

Lazere, Donald. 1992. Teaching the political conflicts: A rhetorical schema. *College Composition and Communication* 43 (2): 194-213.

LeCourt, Donna. 2006. Performing Working-Class Identity in Composition. *College English* 69 (1): 30-51.

Lee. Amy. 2000. *Composing critical pedagogies: Teaching writing as revision.* Urbana, IL: National Council of Teachers of English.

Lindquist, Julie. 1999. Class ethos and the politics of inquiry: What the barroom can teach us about the classroom. *College Composition and Communication* 51 (2): 225-47.

———. 2002. *A place to stand: Politics and persuasion in a working-class bar.* New York: Oxford University Press.

Linkon, Sherry, Peckham, Irvin, and Ben Lanier-Nabors, eds. 2004. *College English* 67 (2).

Lubrano, Alfred. 2004. *Limbo: Blue-collar roots, white collar dreams.* Hoboken, NJ: John Wiley & Sons.

Macedo, Donaldo. 1993. Literacy for stupidification: The pedagogy of big lies. *Harvard Educational Review* 63 (2): 183-206.

MacKenzie, Lawrence. 1998. A pedagogy of respect: Teaching as an ally of working-class college students. In Shepard, McMillan, and Tate 1998, 94-117.

Macrorie, Ken. 1968. *Writing to be read.* New York: Hayden Book Co.

Mack, Nancy. 2000. *Selling academic success to the working class student: Argumentative writing is a failure.* Paper presented at the Conference on College Composition and Communication. Minneapolis, MN. April 12-16.

———. 2006. Ethical representation of working-class lives: Multiple genres, voices, and identities. *Pedagogy* 6 (1), 53-78.

Martin, George T. 1995. My old Kentucky home. In Dews and Law, 1995, 75-85.

Marx, Karl. 1846. *The German Ideology.* http://www.marxists.org/archive/marx/works/1845/german-ideology/ch01b.htm (accessed March 2, 2001).

McLaren, Peter. 1989. On ideology and education: Critical pedagogy and the cultural politics of resistance. In *Critical pedagogy, the state, and cultural struggle,* ed. Henry Giroux and Peter McLaren, 174-204. Albany, NY: State University of New York Press.

McGee, Patrick. 1987. Truth and resistance: Teaching as a form of analysis. *College English* 49 (6): 667-78.

McPeck, John. 1981. *Critical thinking and education.* New York: St. Martin's Press.

Meyrick, Samuel. R. 1800. *A critical inquiry into antient armour.* London: J. Dowding.

Michel, Francisque. 1882. *A critical inquiry into the Scottish language with the view of illustrating the rise and progress of civilisation in Scotland.* London: Blackwood.

Milanes, Cecilia Rodriguez. 1991. Risks, resistance, and rewards: One teacher's story. In Hurlbert and Blitz 1991, 115-24.

Miles, Libby, Michael Pennel, Kim Hensley Owens, Jerekmiah Dyehouse, Helen O'Grady, Nedra Reynolds, Robert Schwegler, and Linda Shamoon. 2008. Interchanges: Commenting on Douglas Downs and Elizabeth Wardle's "Teaching about writing, righting misconceptions." *College Composition and Communication* 59 (3): 503-11.

Miller, Richard. 1998. The Arts of complicity: Pragmatism and the culture of schooling. *College English* 61 (1):10-28.

Moffett, James. 1968. *Teaching the universe of discourse.* New York: Hougton Mifflin.

Mueller, Claus. 1973. *The politics of communication.* New York: Oxford University Press.

National Center for Education Statistics. 2007. *Digest of education statistics.* Learner Outcomes. Table 14-2. http://nces.ed.gov/programs/coe/2008/section2/table.asp?tableID=881 (accessed February 21, 2009).

National Center for Education Statistics. 1999. *Digest of educational statistics tables and figures 1999.* Table 147. http://nces.ed.gov/programs/digest/d99/d99t147.asp (accessed May 27, 2005).

National Center for Education Statistics. 1999. *Digest of educational statistics tables and figures 1999.* Chapter 2, Elementary and Secondary Education. http://nces.ed.gov/programs/digest/d99/lt2.asp (accessed May 27, 2005).

National Center for Education Statistics. 2003. *Digest of educational statistics, 2003.* Table 313. http://nces.ed.gov/programs/digest/d03/tables/dt313.asp (accessed May 27, 2005).

Nevadagrl435. September 22, 2004. The most pathetic book I have ever bought. Review of *Rereading America.* http://www.amazon.com (accessed October 3, 2008).

Newkirk, Thomas. 1989. *Critical thinking and writing: Reclaiming the essay.* Urbana IL: National Council of Teachers of English.

North, Stephen. 1987. *The making of knowledge in composition: Portrait of an emerging field.* Portsmouth, NH: Boynton/Cook Publishers.

Oakes, Jeannie, and Sirotnik, Kenneth. 1983. An immodest proposal: From critical thinking to critical practice for school renewal. ERIC: ED235095.

Ohmann, Richard. 1964. In lieu of a new rhetoric. *College English* 26 (1): 17-22. Quoted in James Berlin 1987. *Rhetoric and reality: Writing instruction in American colleges, 1900-1985.* Carbondale, IL: Southern Illinois University Press.

———. 1982. Reflections on class and language. *College English* 44 (1): 1-17.

Ong, Walter. 2000. *Orality and literacy: The technologizing of the word.* New York: Routledge. (Orig. pub. 1982.)

Parks, Stephen. 2000. *Class politics: The movement for the students' right to their own language.* Urbana, IL: National Council of Teachers of English.

Paul, Richard. 1993. Why students and teachers don't reason well. In *Critical thinking: What every person needs to survive in a rapidly changing world,* ed. Jane Willsen and A. J. Binker. 3rd ed., 151-78. Santa Rosa, CA: Foundation for Critical Thinking. http://outopia.org/teach/resources/CritThink1.pdf (accessed March 3, 2010).

Paul, R. 2004. *Draft statement for national council for excellence in critical thinking.* http://www.criticalthinking.org/about/nationalCouncil.shtml (accessed June 6, 2005).

Peckham, Irvin. 1993. Beyond grades. *Composition Studies/Freshman English News* 21 (2): 16-31.

———. 1997. The yin and yang of genres. In *Genres of writing: Mapping the territories of discourse,* ed. Wendy Bishop and Hans Ostrom, 37-44. Portsmouth, NH: Heinemann.

———. 1999. Whispers from the margin: A class-based interpretation of the conflict between high school and college writing teachers." In *History, reflection and narrative: The professionalization of composition, 1963-1983,* ed. Mary Rosner, Beth Boehm, and Debra Journet, 253-69. Norwood, NJ: Ablex Publishing Co.

———. 2009. Acting justly. In *A 21st century approach to teaching for social justice: Educating for both advocacy & action,* ed. Richard Johnson, III, 75-90. New York: Peter Lang Press.

Plihal, Jane. 1989. Using a critical inquiry perspective to study critical thinking in home economics. *Journal of Vocational Home Economics Education* 7 (1): 36-47.

Pratt, Mary Louise. 1993. Arts of the contact zone. In Bartholomae and Petrosky, eds. 1993, 442-61).

Rhys. Jean. July 19,2007. A great cultural text for freshman composition students. Review of *Rereading America.* http://www.amazon.com (accessed October 3, 2008).

Robinson, Lori. 1992. 'This could have been me': Composition and the implications of cultural perspective. In Berlin and Vivion 1992, 231-43.

Rodriguez, Richard. 1982. *The hunger of memory: The education of Richard Rodriguez.* New York: Bantam Books.

Rosch, Eleanor. 1978. Principles of categorization. In *Cognition and categorization,* ed. Eleanor Rosch and Barbara Lloyd, 27-48. Hillsdale NJ: Lawrence Erlbaum Associates.

Rose, Mike. 1989. *Lives on the boundary.* New York: Penguin Books.

Ryan, Jake, and Charles Sackrey, eds. 1984. *Strangers in paradise: Academics from the working class.* Boston: South End Press.

Sainte-Croix, Guillaume Emmanuel Joseph Guilhem de Clermont-Lodeve, baron de. 1793. *A Critical Inquiry into the life of Alexander the Great, by the ancient historians.* Trans. Richard Clayton. Bath, England: G.G. & J. Robinson.

Scott, Tony. 2009. *Dangerous writing: Understanding the political economy of composition.* Logan: Utah State University Press.

Seitz, David. 1999. Keeping honest: Working class students, difference, and rethinking the critical agenda in composition. In *Under construction: Working at the intersections of composition theory, research, and practice,* ed. Chris Anson Christine Ferris, 65-78. Logan: Utah State University Press.

———. 2004. *Who can afford critical consciousness? Practicing a pedagogy of humility.* Cresskill NJ: Hampton Press.

Shellie. December 2, 2006. One star is too much for this book! Review of *Rereading America.* http://www.amazon.com (accessed October 3, 2008).

Shepard, Alan. 1998. Teaching 'the Renaissance; Queer consciousness and class dysphoria. In Shepard, McMillan, and Tate 1998, 209-30.

Shepard, Alan, John McMillan, and Gary Tate, eds. 1998. *Coming to class.* Portsmouth NY: Boynton/Cook

Shor, Ira. 1992. *Empowering education.* Chicago: University of Chicago Press.

———. 1996. *When students have power: Negotiating authority in a critical pedagogy.* Chicago: University of Chicago Press.

Shor, Ira and Paulo Freire. 1987 *A Pedagogy for liberation: Dialogues on transforming education.* Westport, CT: Bergin & Garvey.

Skolnick, Arlene. 1992. The paradox of perfection. In Colombo, Cullen, and Lisle 1992, 402-09.

Skorozewski, Dawn. 2000. Everybody has their own ideas: Responding to the cliché in student writing. *College Composition and Communication* 52 (2): 220-39.

Slevin, James. 1987. A note on the Wyoming Resolution and ADE. *ADE Bulletin* 87, 50.

Sledd, James. 1991. Why the Wyoming resolution had to be emasculated: A history and its quixoticism. *JAC* 11 (2): 269-81. http://www.jacweb.org/Archived_volumes/Text_articles/V11_I2_Sledd.htm (accessed March 15, 2010).

Smith, Jeff. 1993. "Allan Bloom, Mike Rose, and Paul Goodman." *College English* 55 (7): 721-44.

Smith, Jennifer. September 18, 2007. Necessary! Review of *Rereading America.* http://www.amazon.com (accessed October 8, 2008).

Smith, Othanel. 1953. The improvement of critical thinking. *Progressive Education* 30 (5): 129-134.

Spellmeyer, Kurt. 1993. *Common ground: Dialogue, understanding, and the teaching of composition.* Englewood Cliffs, NJ: Prentice Hall.

Stenberg, Shari. 2006. Liberation theology and liberatory pedagogies. *College English* 68 (3): 271-90.

Strickland, Ronald. 1990. Confrontational pedagogy and traditional literary studies. *College English* 52 (3): 291-300.

———. 1993. Counterstatement. *College Composition and Communication* 44 (2): 250-52.

Stuckey, J. Elspeth. 1991. *The violence of literacy.* Portsmouth NH: Boynton/Cook Publishers.

Sullivan, Patricia. 1998. Passing: A family dissemblance. In Shepard, McMillan, and Tate 1998, 231-51.

Swartz, David. 1997. *The sociology of Pierre Bourdieu.* Chicago: University of Chicago Press.

Tate, Gary. 1998. "Halfway back home." In Shepard, McMillan, and Tate, 252-61.

Tate, Gary, John McMillan, and Elizabeth Woodworth. 1997. Class talk. *Journal of basic writing* 16 (1): 13-26.

Thelin, William. 1993. Counterstatement. *College Composition and Communication* 44 (2): 252-53.

———. 2005. Understanding problems in critical classrooms. *College Composition and Communication* 57 (1): 114-41.

———. 2006. Interchanges: William Thelin's response to Russel Durst. *College Composition and Communication* 58 (1): 114-18.

Thelin, William, and John Paul Tassoni, eds. 2000. *Blundering for a change: Errors and expectations in critical pedagogy.* Portsmouth, NH: Boynton/Cook.

Tompkins, Jane. 1990. Pedagogy of the distressed. *College English* 52 (6): 653-60.

Toulmin, Stephen. 2003. *The uses of argument.* Cambridge: Cambridge University Press. (Orig. pub. 1958.)

Towne, John. 1748. Critical inquiry into the opinions and practice of the ancient philosophers concerning the nature of the soul and a future state. London: Printed for C. Davis.

Trimbur, John. 1989. Consensus and difference in collaborative learning. *College English* 51 (6): 602-27.

———. 1990. John Trimbur's response. *College English* 52 (6): 696-700.

———. 1993a. Articulation theory and the problem of determination: A reading of lives on the boundary. *Journal of Advanced Composition* 13 (1): 33-50.

———. 1993b. Counterstatement. *College Composition and Communication* 44 (3): 248-49.

———. 1994a. The politics of radical pedagogy: A plea for 'a dose of vulgar Marxism'." *College English* 56 (2): 194-206.

———. 1994b. Taking the social turn: Teaching writing post-process. *College Composition and Communication* 45 (1): 108-118.

———. 2000. "Composition and the circulation of writing." *College Composition and Communication* 52 (2): 188-219.

Trusty, Jerry, and Hugh I. Peck. 1994. Achievement, socioeconomic status and self-concepts of fourth-grade students." *Child Study Journal* 24 (4): 281-98.

U. S. Department of Labor, Bureau of Labor Statistics. 2005. *Bureau of Labor Statistics Data.* http://www.bls.gov/data (accessed May 22, 2005).

Villanueva, Victor. 1993. *Bootstraps: From an American academic of color.* Urbana, IL: National Council of Teachers of English.

———. 1998. *An introduction to the social scientific discussions on class.* In Shepard, McMillan, Tate 1998, 262-77.

Vygotsky, Lev Semonovitch. 1975. *Thought and language.* Trans. Eugenia Hanfmann and GertrudeVakar. Cambridge, MA: M.I.T. Press. (Orig. pub. 1962.)

Waite, George. 1846. *A Critical inquiry into a few facts connected with the teeth.* Baltimore: The American Society of Dental Surgeons.

Walford, Geoffrey. 1995. Classifications and framing in English public boarding schools. In *Discourse and reproduction,* ed. Paul Atkinson, Brian Davies, and Sara Deleamont, 191-208. Cresskill, NJ: Hampton Press.

Weiler, Kathleen. 1994. Freire and a feminist pedagogy of difference. In *Politics of liberation: Paths from Freire,* ed. Peter McLaren and Colin Lankshear, 12-40. New York: Routledge.

Whately, Richard. 1963. *The elements of rhetoric.* 7th ed. Ed. Douglas Ehninger. Carbondale, ILL: Southern Illinois University Press. (Orig. pub. 1846.)

Willis, Paul. 1977. *Learning to labor: How working class kids get working class jobs.* Westmead, England: Saxon House.

Wimsatt, W.K., and Monore Beardsley. 1971. The intentional fallacy." In *Critical theory since Plato,* ed. Hazard Adams, 1015-1021. New York: Harcourt Brace Jovanovich.

Wittgenstein, Ludwig. 1953. *Philosophical investigations.* http://www.galilean-library.org/pi7.html (accessed January 10, 2005).

Zandy, Janet, ed. 1990. *Calling home.* New Brunswick, NJ: Rutgers University Press.

———. 1994. *Liberating memory: Our work and our working-class consciousness.* New Brunswick, NJ: Rutger University Press.

Zebroski, James. 1992. The Syracuse writing program and cultural studies: A personal view of the politics of development. In Berlin and Vivion 1992, 87-94.

Zweig, Michael. 2000. *The working class majority.* Ithaca, NY: Cornell University Press.

INDEX

ABOUT THE AUTHOR

IRVIN PECKHAM is the director of the university writing program at Louisiana State University. His two research interests are writing assessment and the intersections of social class and writing instruction. He has published chapters in several edited collections and articles in journals such as *WPA: Writing Program Administration, Composition Studies, Pedagogy, Computers and Composition, English Journal, College Composition and Communication,* and co-edited with Sherry Linkon and Benjamin Lanier-Nabors a special issue of *College English* focusing on social class and writing instruction.